NATIONAL RESEARCH COUNCIL

2101 CONSTITUTION AVENUE WASHINGTON, D. C. 20418

OFFICE OF THE CHAIRMAN

August 14, 1990

The Honorable James D. Watkins
Secretary
U.S. Department of Energy
1000 Independence Avenue, S.W.
Washington, D.C. 20585

Dear Mr. Secretary:

I am pleased to forward "Confronting Climate Change: Strategies for Energy Research and Development," the report of the Committee on Alternative Energy Research and Development Strategies. The report was prepared at the request of Congress and with support from the Department of Energy.

The report was written by a distinguished group of professionals with diverse backgrounds and experience. It provides perspectives on choices and priorities for energy research and development in the United States, outlining ways by which the nation might choose to reduce emissions of greenhouse gases, if it wishes to respond to the possibility of climatic change.

In responding to the request from Congress, the committee was faced with the difficult task of conceiving strategies for the United States while recognizing that actions taken unilaterally by the United States, in the absence of a global strategy involving industrial and industrializing countries, would have limited impact on the global emission of greenhouse gases and could weaken our competitive position in world markets. The committee also had to contend with the reality that the United States and many other nations must rely for a long time on fossil fuels, especially coal. Achieving timely and cost-effective transitions to low- or non-carbon based fuels and energy resources will be very difficult.

Given the magnitude of changes affecting energy production and use that might be required, the committee considered a time horizon that extended to the year 2050. It did this to make distinctions between what might reasonably be accomplished over the longer term and options that could benefit the nation over the next few decades, as well as reflecting heightened concerns about increasing emissions of greenhouse gases. And, given other national priorities and the financial stress in the federal budget, the committee felt compelled to identify actions that were prudent and could be successfully implemented without requiring greater federal appropriations for energy research and development.

The committee also presents an alternative strategy which would require increased investments in energy research and development. It is a strategy to which the nation could commit itself should an increased understanding of the consequences of global warming dictate, or which might be followed to meet other national goals such as energy security.

I believe this report offers insights that are important to the formulation of U.S. energy R&D policy.

Yours sincerely,

Frank Press
Chairman

COMMITTEE ON

ALTERNATIVE ENERGY RESEARCH AND DEVELOPMENT STRATEGIES

DAVID L. MORRISON (Chairman), IIT Research Institute, Chicago, Illinois
JAN BEYEA, National Audubon Society, New York, New York
CLARK W. BULLARD, University of Illinois, Urbana, Illinois
WILLIAM M. BURNETT, Gas Research Institute, Chicago, Illinois
GEORGE M. HIDY, Electric Power Research Institute, Palo Alto, California
HENRY D. JACOBY, Sloan School of Management, Massachusetts Institute of Technology, Cambridge, Massachusetts
EDWARD A. MASON, Amoco Corporation, Chicago, Illinois
JOHN L. MASON, Allied-Signal Aerospace Company, Torrance, California
WILLIAM D. NORDHAUS, Department of Economics, Yale University, New Haven, Connecticut
LESTER P. SILVERMAN,* McKinsey & Company, Inc., Washington, D.C.
HUGH R. WYNNE-EDWARDS,* Moli Energy Limited, Vancouver, British Columbia, Canada

LIAISON WITH THE ENERGY ENGINEERING BOARD

GLENN A. SCHURMANN, Chevron Corporation, San Francisco, California
DAVID C. WHITE, Massachusetts Institute of Technology, Cambridge, Massachusetts

TECHNICAL ADVISERS TO THE COMMITTEE

DEBORAH L. BLEVISS, International Institute for Energy Conservation, Washington, D.C.
DAN STEINMEYER, Monsanto Chemical Company, St. Louis, Missouri
CARL J. WEINBERG, Pacific Gas and Electric Company, San Ramon, California

---

*Lester P. Silverman and Hugh R. Wynne-Edwards served on the committee from June through August 1989 and June through October 1989, respectively, but could not participate in meetings thereafter.

STAFF

Energy Engineering Board

MAHADEVAN (DEV) MANI, Study Director
KAMAL J. ARAJ, Senior Program Officer
ROBERT COHEN, Senior Program Officer
GEORGE LALOS, Consultant
JAMES J. ZUCCHETTO, Senior Program Officer
ARCHIE L. WOOD, Director
PHILOMINA MAMMEN, Administrative Assistant
MARY C. PECHACEK, Administrative Assistant

Building Research Board

PETER H. SMEALLIE, Senior Program Officer

Manufacturing Studies Board

THEODORE W. JONES, Research Associate

EXPERT PANELS

Five expert panels were constituted by the National Research Council to support the Committee on Alternative Energy Research and Development Strategies. Panel members were selected to complement the committee and provide expertise from different sectors of the economy. In the context of the committee's charge directed at reducing emissions of greenhouse gases in energy production and use in the United States, the panels addressed opportunities in the electric power, transportation, buildings, and industry market sectors and strategies for energy R&D and technology adoption. The panels' work was conducted between October 1989 and January 1990, during which time each panel met at least three times. At these meetings the panels invited presentations from and held discussions with representatives of various government agencies (e.g., Departments of Energy and Agriculture, Environmental Protection Agency), from the Oak Ridge, Lawrence Berkeley, Argonne, and Pacific Northwest national laboratories; from the Solar Energy Research Institute; and from industry and academic institutions. The panels subsequently prepared and provided written reports of their respective findings and recommendations as inputs to the committee. These, along with other inputs, were used by the committee in its deliberations and in preparing the final report of the study. However, concurrence of the panels was not sought in the final report. Thus, responsibility for the final report rests solely with the committee. Listings of panels and their membership follow.

ELECTRICITY PANEL
COMMITTEE ON
ALTERNATIVE ENERGY RESEARCH AND DEVELOPMENT STRATEGIES

DAVID C. WHITE (Chairman), Massachusetts Institute of Technology, Cambridge, Massachusetts
THOMAS V. CHEMA, Arter & Hadden, Cleveland, Ohio
OSMAN K. MAWARDI, Collaborative Planners, Inc., Cleveland Heights, Ohio
MOHAN MUNASINGHE, The World Bank/IFC, Washington, D.C.
GUY M. NICHOLS, New England Electric Systems, Westborough, Massachusetts
HEINZ G. PFEIFFER, Consultant, Allentown, Pennsylvania
CARL J. WEINBERG, Pacific Gas and Electric Company, San Ramon, California
ROBERT H. WILLIAMS, Center for Energy and Environmental Studies, Princeton University, Princeton, New Jersey

Staff: Kamal J. Araj, Energy Engineering Board

TRANSPORTATION PANEL
COMMITTEE ON
ALTERNATIVE ENERGY RESEARCH AND DEVELOPMENT STRATEGIES

JOHN L. MASON (Chairman), Allied-Signal Aerospace Company, Torrance, California
HENRY D. JACOBY (Vice-Chairman), Massachusetts Institute of Technology, Cambridge, Massachusetts
CHARLES A. AMANN, General Motors Research Laboratories, Warren, Michigan
DEBORAH L. BLEVISS, International Institute for Energy Conservation, Washington, D.C.
ROBERT J. CASEY, High-Speed Rail Association, Pittsburgh, Pennsylvania
K. G. DULEEP, Energy and Environmental Analysis, Inc., Arlington, Virginia
JACK L. KERREBROCK, Massachusetts Institute of Technology, Cambridge, Massachusetts
RICHARD L. KLIMISCH, General Motors Technical Center, Warren, Michigan
DAVID L. KULP, Ford Motor Company, Dearborn, Michigan
STEPHEN A. PEZDA,* Ford Motor Company, Dearborn, Michigan

Staff: George Lalos, Consultant, Energy Engineering Board

---

*Stephen A. Pezda attended one meeting of the Transportation Panel as an alternate for David L. Kulp.

RESIDENTIAL/COMMERCIAL PANEL
COMMITTEE ON
ALTERNATIVE ENERGY RESEARCH AND DEVELOPMENT STRATEGIES

WILLIAM M. BURNETT (Chairman), Gas Research Institute, Chicago, Illinois
JAN BEYEA (Vice-Chairman), National Audubon Society, New York, New York
JACK B. CHADDOCK, Duke University, Durham, North Carolina
DAVID J. MACFADYEN, NAHB National Research Center, Upper Marlboro, Maryland
DAVID B. GOLDSTEIN, Natural Resources Defense Council, San Francisco, California
MAXINE L. SAVITZ, Garrett Ceramic Component Division, Torrance, California
RICHARD G. STEIN,* Stein Partnership, New York, New York
RICHARD N. WRIGHT, National Institute of Standards and Technology, Gaithersburg, Maryland

Staff: Peter H. Smeallie, Building Research Board

---

*Deceased April 18, 1990.

INDUSTRY PANEL
COMMITTEE ON
ALTERNATIVE ENERGY RESEARCH AND DEVELOPMENT STRATEGIES

DAVID L. MORRISON (Chairman), IIT Research Institute, Chicago, Illinois
JACOB M. GEIST, GeistTec, Allentown, Pennsylvania
EMERY J. HORNYAK, Owens-Illinois, Inc., Toledo, Ohio
NOEL JARRETT, Noel Jarrett Associates, Lower Burell, Pennsylvania
PETER J. KOROS, LTV Steel Company, Independence, Ohio
GEORGE LAUER, Atlantic Richfield Company, Los Angeles, California
MARC H. ROSS, University of Michigan, Ann Arbor, Michigan
DAN STEINMEYER, Monsanto Chemical Company, St. Louis, Missouri

Staff: Robert Cohen, Energy Engineering Board
Theodore W. Jones, Manufacturing Studies Board

R&D STRATEGIES PANEL
COMMITTEE ON
ALTERNATIVE ENERGY RESEARCH AND DEVELOPMENT STRATEGIES

CLARK W. BULLARD (Chairman), University of Illinois, Urbana, Illinois
WILLIAM D. NORDHAUS (Vice-Chairman), Yale University, New Haven, Connecticut
GEORGE M. HIDY, Electric Power Research Institute, Palo Alto, California
RICHARD C. LEVIN, Yale University, New Haven, Connecticut
EGILS MILBERGS, Institute for Illinois, Washington, D.C.
DAVID C. MOWERY, University of California, Berkeley, California
JACK T. SANDERSON, Combustion Engineering, Inc., Stanford, Connecticut
AARON WILDAVSKY, University of California, Berkeley, California

Staff: James J. Zucchetto, Energy Engineering Board

ENERGY ENGINEERING BOARD*

---

*membership Active during the period of study.

ENERGY ENGINEERING BOARD STAFF

ACKNOWLEDGMENTS

The work reflected in this report was performed by the Committee on Alternative Energy Research and Development Strategies between June 1989 and February 1990. The committee wishes to acknowledge with gratitude the assistance of the following individuals who provided information on various topics of interest during the course of the study:

The Honorable Timothy E. Wirth, United States Senate (Colorado); David Harwood, Staff, U.S. Senate; John Ahearne, Society of Sigma Xi; Marvin Garfinkel and Walter L. Robb, General Electric Company; D. Warner North, Decision Focus, Inc.; Bob Jones, ADM Milling Company; J. Laurence Kulp, Consultant; William Pepper, ICF, Inc.; Geoffrey J. Sturgess, Pratt & Whitney, United Technologies Corporation; John Doyle, Pacific Gas and Electric Company; John F. Elliott and Nicholas J. Grant, Massachusetts Institute of Technology; David Pimentel, Cornell University; Ronald B. Edelstein and Thomas Hayes, Gas Research Institute; Jeremy Metz, American Paper Institute; Eric Vaughn, Renewable Fuels Association; Nicholas Fedoruk, Environmental Action Foundation; John O. Berga and Timothy S. Yau, Electric Power Research Institute; David Jhirad, U.S. Agency for International Development; Frederick M. Bernthal, U.S. Department of State; Patrick R. Booher, Richard A. Bradley, Paul J. Brown, Thomas A. Calhourn, Albert A. Chesnes, John F. Clarke, Melvin H. Chiogioji, Mary Corrigan, David H. Crandall, Bruce Cranford, J. Michael Davis, George Doumani, John N. Eustis, Kenneth M. Friedman, Thomas J. Gross, Robert Kane, David J. McGoff, David O. Moses, David B. Nelson, Robert Rabson, David M. Richman, Howard Rohm, Peter H. Salmon-Cox, Robert L. San Martin, Alan J. Streb, Linda G. Stuntz, Denise F. Swink, Donald K. Walter, and Edward R. Williams, U.S. Department of Energy; Earl Gavett and Norton D. Strommen, U.S. Department of Agriculture; Robert Flaak, Richard Morgenstern, Barry Solomon, and Dennis Tirpak, U.S. Environmental Protection Agency; Linda Berry, Roger Carlsmith, Robin Cantor, Phil Fairchild, William Fulkerson, Ed Hillsman, Eric Hirst, Michael Kuliasha, Jack Ranney, John Reed, David Reister, Marty Schweizter, Robert Van Hook, and Robert Wendt, Oak Ridge National Laboratory; Brandt Anderson, Sam Berman, William Carroll, Joan Daisey, Rick Diamond, Don Grether, Ashok Gadgil, Arlon Hunt, Joseph Klems, Jon Koomey, Mark Levine, Arthur Rosenfeld, Michael Rubin, Steve Selkowitz, Michael Wahlig, and Fred Winkelmann, Lawrence Berkeley Laboratory; Don M. Rote, Argonne National Laboratory; W. Bradford Ashton and Jae A. Edmonds, Battelle Pacific Northwest Laboratory; Stanley Bull, Solar Energy Research Institute; John Facey, National Aeronautics and Space Administration; David L. Bodde, Robert A. Coppock, Andrew C. Lemer, and Lawrence E. McCray, National Research Council.

CONTENTS

List of Tables

List of Figures

HIGHLIGHTS

Human activities, primarily related to energy production and consumption, are increasing concentrations of greenhouse gases (GHGs) in the atmosphere faster than the rate of absorption by natural sinks. If this continues, there is concern that irreversible climatic changes may occur and adversely impact human societies. However, scientific uncertainty exists regarding the timing and extent of global climate change and its consequences.

The committee was asked, nevertheless, to suggest federal strategies for energy research and development (R&D) on alternative energy technologies should the government want to give priority to stabilizing atmospheric concentrations of GHGs. In response, the committee has recommended strategies to facilitate prudent and decisive actions by the United States.

1. There is no immediate single technological fix for reduction of GHG emissions. Consequently, if GHG emissions are to be reduced, a multifaceted effort will be required by the U.S. government and the private sector, on major improvements in U.S. energy productivity, on the adoption of relevant technologies in the marketplace, on energy R&D, and on the resolution of environmental and economic issues.

2. As a first step and within existing budget constraints, changes to U.S. energy R&D priorities are warranted whose implementation would

(a) enable reductions in GHG emissions and
(b) merit public expenditures for other reasons, even in the absence of concerns over global climate change.

The underlying R&D strategy (referred to as the Focused R&D Strategy) has the objective of reducing uncertainties about the costs and performance of low- or non-GHG-emitting technologies relative to their conventional counterparts.

3. From a programmatic standpoint consistent with the foregoing strategy, the U.S. Department of Energy (DOE) should take the following steps as soon as possible:

(a) Allocate to energy programs on conservation and renewables an additional $300 million or about 20 percent of the civilian energy R&D budget appropriations
(b) Obtain the funds for this allocation by refocusing the R&D priorities for magnetic fusion and fossil energy programs

(c) Redirect efforts under the fossil energy program to achieve high efficiencies in converting clean coal to electricity

(d) Make international cooperation an important component of alternative energy R&D projects.

4. In the near term (i.e., from the year 1990 to 2000), and as a supplement to the Focused R&D Strategy, the government ought to consider taking additional actions to stimulate

(a) adoption of GHG-reducing technologies that already exist but that are not currently deployed

(b) private sector investments in energy R&D on low- or non-GHG technologies.

5. If concerns heighten regarding GHG emissions, an additional R&D strategy (referred to as the Insurance Strategy) should be considered to develop and demonstrate, for their "insurance" value, "backstop" technologies to reduce GHG emissions. Substantial federal funding (on the order of billions of dollars per year) of applied energy research, development, and demonstration would be needed to minimize uncertainties in the economic and environmental costs of technologies that reduce GHG emissions. Major national policy decisions are required to adopt the Insurance Strategy.

6. The U.S. government must continue to emphasize basic and generic research on many fronts to expand knowledge of alternative energy sources.

7. The United States ought to take the lead in GHG-reducing energy R&D in an international context.

1

STRATEGIES FOR REDUCING GREENHOUSE GAS EMISSIONS: KEY FINDINGS AND RECOMMENDATIONS

Emissions of greenhouse gases (GHGs), especially from energy production and use, and their impact on global climate emerged as a major national issue in the United States during the 1980s. As a result, Congress (P.L. 100-371, 1988) directed the U.S. Department of Energy (DOE) to ask the National Academy of Sciences and the National Academy of Engineering to assess the current state of research and development (R&D) in the United States in alternative energy sources, and to suggest energy R&D strategies involving roles for both the public and private sectors, should the government want to give priority to stabilizing atmospheric concentrations of GHGs.

The findings and recommendations of the Committee on Alternative Energy Research and Development Strategies, appointed by the National Research Council in response to Congress's directive, are provided in this report and summarized in this chapter. The energy R&D strategies and actions recommended by the committee are structured to facilitate prudent and decisive responses by the United States, despite uncertainties regarding the effects of GHGs on global climate.

END-USE SECTOR ANALYSIS

Electric Power Sector

The electric power sector has the potential to produce and deliver electricity essentially free of GHG emissions, primarily CO_2. Currently, however, electricity is generated worldwide predominantly from fossil fuels, with coal being the dominant fuel choice. Low- or non-CO_2-emitting power generation technologies based on nuclear fission reactors, renewable resources, and geothermal energy are commercially available, and technically, could supply the world's energy needs. Because of unfavorable economics as well as environmental, health, and safety concerns, it is by no means clear that these technologies could be deployed on the scale required without substantial research, development, and demonstration (RD&D) and lower costs. The transition from coal to these low- or non-CO_2-emitting technologies will involve major changes in the selection and deployment of generating facilities in energy markets served by the electric utilities and in the economy as a whole. Integral to the transition are efficiency improvements in generation; installation of cogenerating units; shifts or retrofits of generation capacity to lower carbon fuels, such as natural gas; and improvements in transmission, distribution, and storage systems to reduce energy losses and maximize use of the most efficient generating facilities.

Transportation Sector

In the United States cars and light trucks produce well over half the GHG emissions attributed to the transportation sector. The most significant near-term opportunities for reducing these emissions are to be found in improving the fuel efficiency of cars and light trucks and in using transportation systems more efficiently (e.g., increasing the average load factor of passenger cars). Generally, realizing these opportunities would not require new technology, but rather would exploit existing technology that is ready (or nearly ready) for commercial application. If, in the long term, major reductions in GHG emissions must be achieved, even higher energy productivity increases must be attained and new energy sources and fuels must be found.

Innovation in cars and light trucks is almost completely in the hands of the large domestic and foreign manufacturers, most of whom perform extensive R&D on new products. The federal role in technology R&D should be limited to strengthening the technology base and nurturing research in potentially important areas in which the automotive companies are not likely to invest heavily (e.g., basic research on materials and advanced batteries). Public policy must be shaped to stimulate additional improvements in vehicle efficiencies and more efficient use of transportation systems if GHGs are to be reduced.

Buildings Sector

Over the long term, energy use and GHG emissions in residential and commercial buildings can be reduced by more than 70 percent through successful development and implementation of technology. Energy demand in buildings can be decreased by technologies that reduce heat loss, such as improved wall, window, and roofing materials and insulation. The demands can be further decreased by increasing the efficiency of the equipment used to heat, cool, ventilate, and light buildings. Also, high-efficiency appliances and office equipment can be developed to reduce energy use. Finally, technological opportunities exist for greater use of cogeneration systems, natural gas, CO_2 neutral fuels derived from biomass, and non-GHG energy resources (such as solar) to reduce GHG emissions. These opportunities must be pursued by a combination of private and federal R&D accompanied by public policies and incentives for technology implementation, because the number of decision makers involved in energy-related issues in the buildings sector approaches the total population of building users from homeowners and tenants to shopkeepers, office workers, and building managers.

Industry Sector

The efficiency of energy use can be improved and the form of energy use can be altered to significantly reduce industrial emissions of GHGs. In many instances the technologies are currently available, but neither market signals nor government policies encourage investments to achieve these improvements. Continued R&D efforts by industry to improve processes and reduce energy use, complemented by collaborative projects with the federal government and the results of basic research, will reduce the energy required per unit of production by an average of about 1.5 percent per year. Changing fuels can reduce GHG emissions, but implementation depends on the relative price and availability of electricity and natural gas versus coal and oil. Increased recycling of materials also offers a means of reducing energy use and GHG emissions.

CURRENT STATUS OF ALTERNATIVE ENERGY R&D

Through its assessment of alternative energy R&D programs within the federal government, it is apparent to the committee that only limited investments are being made in technologies relevant to the reduction of GHG emissions. For example, federal funding for DOE's civilian energy R&D in the solar and renewables program declined by 89 percent on a constant dollar basis, from FY 1979 to FY 1989; conservation program funding declined by 61 percent, electric energy program funding declined by 76 percent, and funding for the nuclear fission program declined by 78 percent. Private sector funding by individual companies also declined during the 1980s for conservation and renewable energy technologies. The Gas Research Institute and the Electric Power Research Institute have R&D programs in end-use technologies and conservation and in renewable energy that are complementary to DOE's efforts. In the aggregate, however, current funding for alternative energy R&D in the United States is not sufficient to address the problem of achieving major reductions in GHG emissions.

FINDINGS

No single technological fix that would significantly reduce GHG emissions during the next few decades was identified by the committee in any of the four end-use sectors. The uses of energy are too diverse. Rather, two broad technological pathways exist that by the year 2050 could lead to significant reductions (from today's levels) in GHG emissions. These pathways, which are not mutually exclusive, involve

- Increases in energy productivity through improvements in the efficiency of energy use and conversion technologies and
- Development of and shift to the use of low- or non-GHG-emitting energy technologies.

The pace at which the nation can pursue either or both of these pathways must be tempered by prevailing international economic competitiveness and by issues related to domestic energy supplies and the environment. If the United States can become highly efficient in the production and use of energy, the burden of achieving a shift to non-GHG-emitting alternative energy sources would be greatly reduced.

Up to the year 2000, the only technologies that can have a significant impact on the reduction of GHG emissions are those already developed and available. While R&D per se will have little effect on the adoption of these technologies, incentives and regulations by the federal government can have major influence. Changes in public policies will be required if markets are to be stimulated to adopt available technologies that are highly energy efficient.

RECOMMENDATIONS

The following two energy R&D strategies, which could lay the groundwork for achieving reductions in GHG emissions, are recommended to the federal government:

- Focused R&D Strategy. Pursue energy R&D that is aimed at reducing GHG emissions and that would make sense for other reasons even in the absence of concerns about global climate change.

- Insurance Strategy. Pursue energy R&D that would be viable only in the presence of concerns about global climate change.

Each R&D strategy addresses, in general terms roughly the same set of technologies and spans the full range of activities from fundamental research to technology adoption but with differing objectives and costs.

Both strategies follow the conventional R&D paradigm of reducing uncertainties about the cost and performance of a technology by producing new knowledge. The fundamental difference between them is the difference in the magnitude, timing, and costs of actions that can be justified on non-GHG grounds and those that need a GHG justification. In both strategies federal funding is needed because of the inability of private firms to capture the benefits of basic research and because the price of fossil fuels is less than the full social cost associated with their use.

The federal R&D program under the Insurance Strategy will be considerably more costly to the government (involving multibillion dollar increments over the Focused R&D Strategy), and a greater fraction of the government's R&D would be directed toward reducing the uncertainties associated with the technology-adoption phase. Through the Insurance Strategy the nation would, over time, invest in the development and demonstration of a variety of "backstop" technologies for their "insurance" or option value.

It is not sufficient to define energy R&D priorities in isolation from the marketplace for which the products of the R&D are intended. Prevailing market forces must be considered and government actions may be required to achieve specific national objectives. In the past, particularly at times of crisis, the government has used intervention mechanisms such as taxes, tax credits, energy efficiency standards, loan guarantees, subsidies, federal procurements, and liability limitations to influence the supply and demand of fuels and energy resources. In the event that the nation makes a commitment to reduce emissions of GHGs significantly, such actions ought to be considered again as a supplement to the Focused R&D and Insurance strategies. This would stimulate energy R&D in the private sector and the adoption of GHG-reducing technologies in the marketplace.

In the near term (i.e., from the year 1990 to 2000), such actions could spur the adoption of GHG-reducing technologies that already exist and that can be shown to be economically viable for reasons other than low-GHG emissions but that are not currently being used. Similarly, in the longer term such actions could apply to low- or non-GHG-emitting technologies for which R&D has helped reduce cost and performance uncertainties. The committee has not evaluated the efficacy of market intervention mechanisms that might be appropriate for achieving various levels of reductions in GHG emissions over time, but the evaluations ought to be done before the government invokes such actions.

Focused R&D Strategy - Actions

The high-priority energy R&D opportunities and enabling policies that ought to be addressed now in the context of this strategy are highlighted below. The actions suggested represent those considered to be the most important from among a longer list of promising options analyzed by the committee. Their execution entails changes to the current federal R&D program priorities and selective reprogramming of R&D funds within existing budget outlays.

- Fossil Energy

  Increase the efficiency of electricity generation using currently available high-efficiency options such as the gas turbine/steam turbine combined cycle

  Develop substantial improvements in the combined cycle and other advanced gas-turbine-based technologies for firing with natural gas or a gaseous fuel derived from biomass

  Achieve economic recovery of gas from known domestic reserves

Improve reservoir characterization through basic geoscience research to enable future resource recovery

Define GHG emissions as one criterion in evaluating new approaches to coal combustion

- Nuclear Energy

Determine through social science research the conditions under which nuclear options would be publicly acceptable in the United States

Conduct an international study to establish criteria for globally acceptable nuclear reactors

- Conservation and Renewable Energy

Utility Systems

Provide RD&D support to new and improved technology for electric storage and for alternating current and direct current systems components

Develop an efficient, flexible, and reliable network to operate the electric power system in the most environmentally acceptable way

Photovoltaics

Accelerate R&D on materials and module manufacturing to increase efficiency and reduce costs of photovoltaic systems

Transportation Technologies

Improve batteries for vehicle propulsion to achieve higher performance and durability and reduce costs

Adapt alternative fuels (e.g., alcohols) to engines and vehicles

Reduce emissions from efficient power plants such as the diesel

Evaluate vehicle systems to assure the safety of smaller cars built with lightweight structural materials

Investigate innovative electric transportation systems

Building Envelope/Superinsulation

Develop advanced insulation materials for building walls, windows, and roofs

Develop non-GHG foams and evacuated panel technology

Building Operating Practice

Develop controls, expert systems, diagnostics, and feedback systems to minimize energy use in the construction, commissioning, and operation of buildings

Building Implementation R&D

Implement existing technologies with carefully planned and monitored demonstrations and research on motivation and decision making

Industrial Process Energy Efficiency

Continue industry-government cooperative programs such as the metals initiative (i.e., steel and aluminum)

Recycling of Materials

Develop improved separation technologies

Create markets for postconsumer-recycled materials in the manufacture of high-quality products

Biomass and Biofuel Systems

Expand through basic research, our understanding of the mechanisms of photosynthesis and genetic factors that influence plant growth

Perform systems analyses to define and prioritize infrastructure requirements with expanded use of biomass-derived fuels

Assess the potential environmental impacts of biomass production (e.g., through silviculture), including impacts on biodiversity and the availability of water resources

The R&D activities outlined above would have to be supplemented by government actions to stimulate the adoption of technologies and processes for reducing GHG emissions. In the near term such actions would include the following:

- Electric Power

  Increase U.S. nuclear power plant availabilities to levels conforming to the best international practice

  Facilitate greater environmental dispatch of generation facilities

- Transportation

  Inform consumers and develop and implement policies to stimulate the market for cars and light trucks that are significantly more energy efficient than current models

  Develop and implement policies to achieve higher productivity of transportation energy use

- Buildings

  Enact substantial changes in the regulatory environment to allow electric and gas utilities to earn from investments in energy productivity as well as energy supply and to decouple utilities' net revenues from their sales volumes

  Stimulate, through competitive bidding, nonutility investments in energy supply and conservation that reduce GHG emissions

- Industry

  Encourage front-end separation of wastes through incentives or penalties

  Eliminate regulations counterproductive to waste management and recycling

  Encourage, in conjunction with the electric power sector, the installation of cogeneration units

Insurance Strategy

This strategy incorporates the lessons of past failures with large R&D projects and envisions major outlays of federal funds to develop and demonstrate the viability of promising low- or non-GHG-emitting technology options for "insurance" purposes. The RD&D would be undertaken even though the technologies are clearly not cost competitive today in comparison to their higher-GHG-emitting, fossil-fueled counterparts (and may never be feasible without federal support of R&D and market intervention). These alternative technologies would be developed to the level necessary to understand their costs and impacts if concerns about GHG emissions and climate change heighten and the need to deploy such technologies in the marketplace becomes much more compelling.

Because of their multibillion dollar cost implications, decisions regarding the Insurance Strategy and related market interventions ought to be made in light of other national and international policy considerations. The committee has identified energy RD&D targets that would be important to pursue under this strategy. The choice of policy instruments that the government might consider for market intervention to accompany the RD&D would depend on the magnitude and timing of GHG reductions to be achieved. The choice would also be strongly influenced by the difference between the prevailing price of carbon-based energy and the cost of the new technology for displacing it.

- Fossil Energy

  Fund an exploratory study to ascertain if there are viable approaches (economically and environmentally) for removing and sequestering CO_2

- Nuclear Energy

  On the strength of the public acceptability and global reactor studies performed under the Focused R&D Strategy, fund an industry-led or -managed program to develop and demonstrate an advanced reactor

- Conservation and Renewable Energy

  Stimulate production (at the rate of about 10 megawatts per year each) of the three to five most promising photovoltaic technologies; the same should be done in the areas of solar thermal and wind energy conversion

  Demonstrate "new" projected storage systems such as compressed gas, battery arrays, and superconducting magnets

  Develop approaches for federal cost-sharing and utility procurements of renewable energy technologies or electricity generated by them. Such financing mechanisms should enable manufacturers to compete in niche markets (both domestic and export) to sustain production at levels sufficient to determine the ultimate potential of the technologies.

  Select a major metropolitan center at which to demonstrate higher productivity of transportation energy use

Demonstrate the efficacy of an electric transportation system in at least one major city

Develop and demonstrate photovoltaic electricity resources for buildings, including lighting and water heating

Develop and demonstrate advanced design, construction, and management practices in programs involving utilities, building authorities of local governments, and energy service companies

Reduce energy use in existing buildings through adoption of insulation retrofits, window replacement, and intensive use of diagnostic technologies over a 10-year period

- Recycling of Materials

  Conduct a major demonstration program to determine the feasibility of greatly increased recycling in several industrial processes

- Biomass and Biofuels System

  Develop and demonstrate promising biomass-to-fuels conversion processes, particularly for cellulose and hemicellulose

  Select and demonstrate on a large scale the use of improved plant species to enhance biomass production

  Develop strategies to mitigate environmental impacts of large-scale use of biomass

Basic and Generic Research

The key to realizing the promising technological opportunities for significantly reducing GHG emissions in energy production and use is to perform the underlying basic and generic research. Fundamental research to expand the knowledge base of science and engineering relevant to fuels, materials, processes, and energy systems will facilitate the development of technologies under any R&D strategy the nation may choose to pursue. Particularly important is research in areas such as materials (high-temperature, lightweight, structural); plant physiology, biochemistry, and genetics; energy conversion devices and systems; and social, behavioral, and environmental sciences. Such research must be assiduously nurtured.

GHG Emissions Monitoring and Instrumentation

The nation is currently ill equipped to correctly formulate and effectively manage an alternative energy R&D strategy because of immense uncertainties surrounding GHG emissions and global climate change. Although an adequate understanding of these relationships could be extracted from information already deposited in the paleogeological record, fundamental scientific uncertainties are likely to persist for decades. Because of the inherently long response times of the phenomena involved, it will take time to test hypotheses and build the necessary knowledge base for a better understanding of global climate change. The United States should lead in an international program to achieve the needed monitoring instrumentation and improve climate modeling capability.

Federal Outlays for Alternative Energy R&D

The high priority R&D initiatives and technology-adoption actions described under the Focused R&D Strategy are estimated to require an incremental annual and sustained funding level equal to about 20 percent of the 1990 civilian energy R&D budget, or approximately $300 million (in 1990 dollars) per year. An approximate distribution of these incremental funds should be as follows: electrical storage, $15 million; photovoltaics, $30 million; biomass, $60 million; buildings, $105 million; recycling, $75 million; and transportation research, $15 million. The committee recommends that DOE initially obtain these funds by reprogramming its efforts in the fusion, fossil energy, and other programs and reallocating them to R&D in conservation and renewables. To the extent possible, funds currently budgeted for the clean coal technology program and for the civilian nuclear reactor development program should be reallocated within those programs to achieve, respectively, high conversion efficiencies of coal to electricity and, with international collaboration, the definition of criteria for globally acceptable reactors. Clearly, international cooperative efforts will almost certainly be required on a number of future technological options such as biomass and renewables, and the proposed nuclear study could exemplify how such undertakings should be planned and conducted. The committee finds it highly unlikely that commercially viable magnetic fusion reactors will make additions of any significance to the U.S. electricity generation mix before the year 2050. Hence, the emphasis of magnetic fusion R&D in the United States should be on basic research and greater international collaboration.

No definitive estimate was made for the funding required for the Insurance Strategy. Depending on the scope of the development programs undertaken and the number of demonstration projects initiated to facilitate the adoption of the GHG-stabilizing technologies, the magnitude of federal expenditures could range from $100 million to $500 million per year for up to 10 years in each of the end-use sectors and for each major technology option to generate electricity free of GHG emissions. Such federal expenditures would have to be accompanied by private sector funding if programmatic goals and technology adoption are to be achieved.

Leveraging Federal Investments Globally

The global character of the GHG issue imposes a special requirement on R&D. Both the advancement of science and the development of alternate "solutions" require an international context. The foreseeable R&D costs to make progress will be high in these two areas; hence, it would be desirable to share these costs as broadly as possible. A major opportunity is at hand for RD&D in cooperation with the developed countries to seek options for energy supply in the developing world. International cooperation in energy RD&D can be encouraged through governmental arrangements and by ad hoc agreements with energy producers. Without international cooperation to stabilize GHG emissions, the efforts by any single nation will fall far short of global needs.

2

BACKGROUND

GENESIS OF THE STUDY

Global climate change emerged as a major issue in the United States during the 1980s. As a result of national concern, Congress adopted legislation in its 1989 Energy and Water Development Appropriation Bill for studies to clarify and define the extent of the problem and to recommend options for its management. The legislation was based on the premise that emissions of greenhouse gases (GHGs) arising from energy production and use are significant precursors of global climate change. Accordingly, in Public Law 100-371, Congress directed the secretary of energy to ask the National Academy of Sciences and the National Academy of Engineering to:

- assess the current state of research and development (R&D) in the United States in alternative energy sources "including, but not limited to, nuclear power, solar power, renewable energy sources, improved methods of employing fossil fuels, energy conservation, and energy efficiency";

- suggest R&D strategies for stabilizing the atmospheric concentrations of GHGs that contribute to global climate change; and

- analyze what federal investments would encourage greater private investment in alternative sources of energy.

This report presents the recommendations of the study committee (Committee on Alternative Energy Research and Development Strategies) appointed by the National Research Council in response to the directive.

PROBLEM DESCRIPTION

A number of atmospheric gases absorb infrared radiation emitted from the earth's surface and prevent its escape into space. This trapping of infrared radiation is commonly referred to as the greenhouse effect. The principal GHGs that are also constituents of the atmosphere are carbon dioxide (CO_2), water vapor (H_2O), and methane (CH_4). Other important GHGs include ozone, nitrous oxide (N_2O), and the chlorofluorocarbons (CFCs); of these the CFCs have no natural sources. Altogether, over 40 GHGs have been identified so far, most of which are radiatively active.[1-3]

GHGs must be present in the earth's atmosphere for the earth's temperature to be suitable for life as we know it. However, human activities (primarily energy-related ones, as shown in Figure 2-1) are increasing the atmospheric concentrations of many GHGs at a rate that is faster than the rate of absorption by natural sinks. The concern is that, if this rate of increase in the concentrations of GHGs continues, climatic changes may arise that would have major impacts on the natural environment and on human societies.[4-9] Scientific uncertainty exists, however, regarding the timing and extent of potential global climate change from the accumulation of GHGs in the atmosphere.[10,11]

Principal GHGs associated with energy production and use include CO_2 emitted during the combustion of hydrocarbon fuels; CH_4 emissions from coal mines and from the venting and leakage of natural gas during drilling, production, and transmission; and releases of CFCs from air conditioners, refrigeration equipment, and the production of insulating materials. N_2O emissions come from the combustion of hydrocarbon fuels, including agricultural biomass, and from the use of nitrogenous fertilizers.

The atmospheric concentration of CO_2 is currently around 350 parts per million by volume (ppmv), which is about 25 percent above the preindustrial level of 280 $\pm$ 10 ppm estimated for 1860 (Figure 2-2) and exceeds atmospheric concentrations that can be inferred from geologic records.[13]

On average, CO_2 emissions from fossil fuel combustion have increased by 4.3 percent per year since 1860 and are currently on the order of 20 billion metric tons of CO_2 or 5.5 billion metric tons of carbon (GTC) per year out of a total of 6 to 8 GTC per year from all man-made sources. The contribution to CO_2 emissions of hydrocarbons used for fuel depends on the carbon content of the fuel per unit of combustion energy. Among the fossil fuels, per quad (10^{15} Btu) of energy used, natural gas emits the least carbon (14.5 million metric tons carbon, MTC); coal the most (25.1 MTC); and petroleum an intermediate amount (20.3 MTC). Traditional biomass fuels, such as crop residues and dung, release CO_2 to the atmosphere in a balanced cycle of absorption and respiration. In contrast, other biomass fuels such as firewood may provide either a net annual source or sink for carbon, depending on whether the underlying biomass stock is being depleted or increased, respectively.

Although the concentrations by volume and the annual rate of emissions of other GHGs such as CH_4, the CFCs, and N_2O are much less than those of CO_2, they cannot be ignored (Table 2-1). Carbon dioxide is nevertheless the single most significant anthropogenic GHG from energy production and use (Figure 2-3), and its control was the focus of this study.

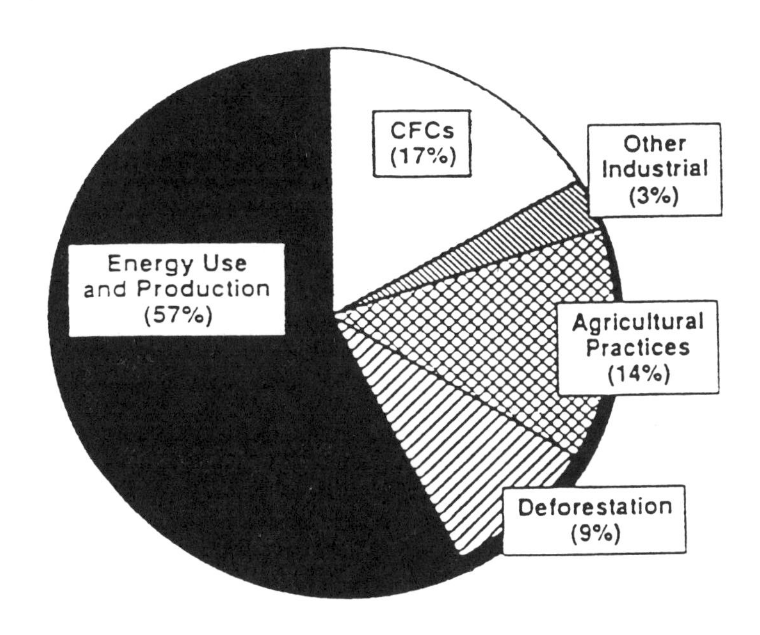

FIGURE 2-1 Sources of GHGs. Energy production and use constitute the largest human source of greenhouse gases, but other activities are also significant.[2]

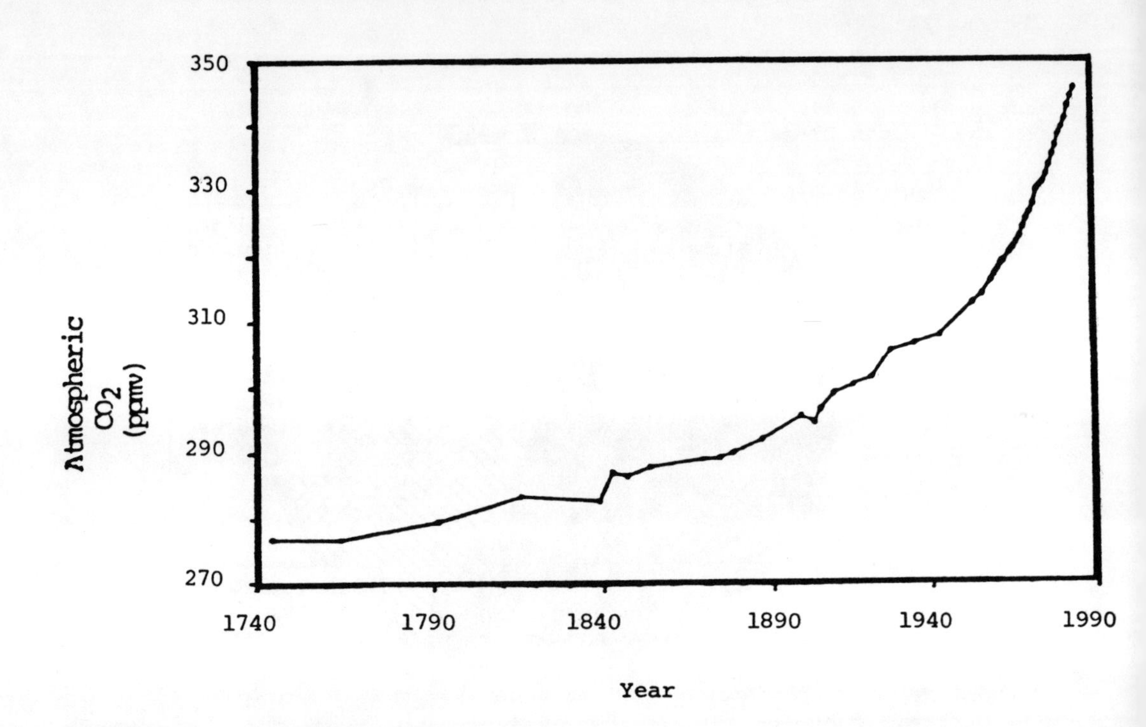

FIGURE 2-2 Historical variation in atmospheric carbon dioxide concentration.[12]

TABLE 2-1 Key Atmospheric Trace Gases Whose Concentrations Are Increasing[14]

Gas	Concentration[a] in 1985	Annual Rate of Increase As of 1985
CO_2	345 ppmv	1.4 ppmv (0.4%)
CH_4	1.65 ppmv	18.0 ppbv (1.1%)
N_2O	305 ppbv	0.6 ppbv (0.2%)
CFC-11	220 pptv	11.0 pptv (5.0%)[b]
CFC-12	380 pptv	19.0 pptv (5.0%)[b]

[a] Concentrations are global averages in 1985. Values shown for the rate of increase are representative as of 1985.

[b] These chlorofluorocarbons will be phased out under the terms of the Montreal Protocol that is now in force.

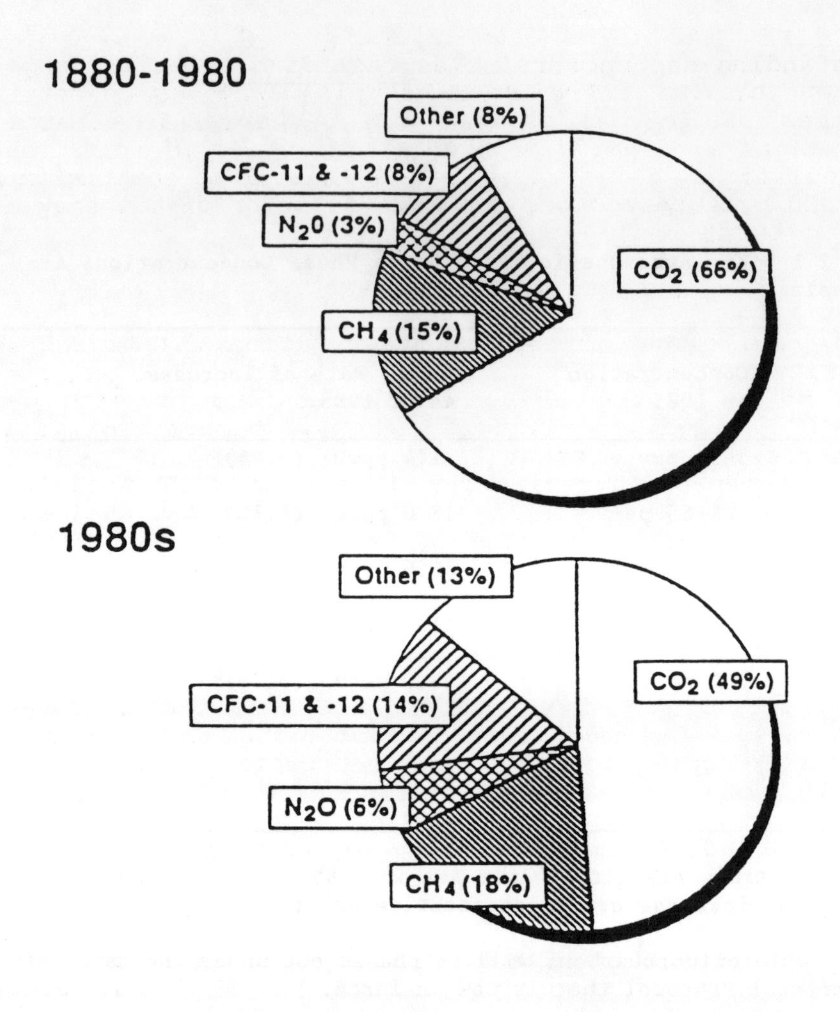

FIGURE 2-3 GHGs responsible for increases in the greenhouse effect worldwide.[2]

Notwithstanding the uncertainties regarding climate change and its consequences, the central task of this study was to determine the priorities and federal strategies for energy R&D efforts and the deployment of alternative energy technologies to significantly reduce GHG emissions. The strategies are to include actions in both the public and private sectors and consideration of how they might mesh and complement one another.

THE GLOBAL CONTEXT

On the premise that increasing accumulation of GHGs increases the probability that significant global warming will occur, a major goal would be to reduce atmospheric emissions of such gases, especially CO_2. Global fossil fuel energy resources are large and include petroleum, coal, natural gas, tar sands, oil shale, and deposits of bitumen.[15] Approximately 80 percent of the world's coal resources are in the United States, the U.S.S.R., and the People's Republic of China, and coal is expected to be the dominant fuel that will be used around the world over the next several decades. Continued use of increasing amounts of fossil fuels, unconstrained by considerations of the potential impacts of GHG emissions, could lead to atmospheric accumulation of CO_2 approaching concentrations likely to initiate irreversible changes in the earth's climate. Controls may therefore be required on the use of such fuels and could be targeted to hold CO_2 concentrations and the rate of increase from exceeding some generally acceptable limits.

The U.S. Environmental Protection Agency's study, Policy Options for Stabilizing Global Climate,[2] concluded that very large reductions (on the order of 50 to 80 percent of current levels) in worldwide CO_2 emissions are required, starting now, to achieve stabilization of atmospheric GHG concentrations at their current levels. Despite considerable uncertainties in this estimate, it raises at least two fundamental questions: (1) How much reduction in the emissions of GHGs (from a specified baseline) is achievable with various energy production and end-use technologies? (2) What would it take to implement those technologies (whether currently available or yet to be developed) and effectively replace their less energy-inefficient or more polluting counterparts? As a first step, this study addresses these questions for the electric power, transportation, buildings, and industry sectors in the United States.

Aside from the CFCs (phaseout of which is now governed by the Montreal Protocol), increases in worldwide emissions of CO_2 and other GHGs are expected to continue over the next century unless strong public policies are adopted for their control. It is further expected that the bulk of the emissions will result from increasing exploitation of hydrocarbon fuels, primarily coal, necessitated by energy demands from industrialized countries such as the United States and the U.S.S.R., from countries with expanding industrial development such as China and India, and from growing populations. Concerted action by all countries is essential if successful responses to regional and global environmental problems such as acid rain and the greenhouse effect are to be developed and implemented in a timely and effective manner. What is still an open issue is how such action is to be taken worldwide and what the roles and responsibilities of various countries would be to assure its successful execution.

For developing countries, all of which will need more energy to fuel their economic growth, even the prospect of restrictions on the use of relatively cheap, easily accessible fossil fuels raises questions of equity and fairness. Their view is that the bulk of the burden of reducing the use of fossil fuels ought to rest on the industrialized countries, because annual consumption of energy resources in those countries has thus far accounted for about 80 percent of the world's consumption and attendant emissions of GHGs.

The United States is currently the primary contributor to the greenhouse effect. According to a recent study,[16] U.S. emissions of CO_2 have increased over the past 2 years, and the U.S. share of global CO_2 emissions in 1988 was estimated to be around one-quarter. In this milieu there is growing opinion in the United States that a number of innovative and cost-effective U.S. actions are possible, which could significantly reduce GHG emissions from the nation's current levels, establish a leadership position for the United States with which to support other countries in similar efforts, and create a setting in which actions needed worldwide can be planned and executed in a timely and concerted manner.[17-19]

The committee appreciates the value in such a role being taken by the United States even as it recognizes that unilateral action taken solely by one country will be much less effective than concerted actions. In the committee's view, developing a strategic vision for the United States in terms of energy R&D and adoption of alternative energy technologies that are low or even nonemitters of GHGs is an important first step. The vision should then serve and be used as a point of departure toward a broader program of global cooperation and joint efforts for safeguarding the environment. This study was approached with such a viewpoint and expectation.

NOTES AND REFERENCES

1. See, for example, A Primer on Greenhouse Gases: CO_2, Report No. TR040, DOE/NBB 0083, U.S. Department of Energy, Office of Energy Research, Office of Basic Energy Sciences, Carbon Dioxide Research Division, Washington, D.C., March 1988.

2. U.S. Environmental Protection Agency, Office of Policy, Planning and Evaluation, presentation to the Committee on Alternative Energy Research and Development Strategies, National Research Council, Washington, D.C., June 12, 1989.

3. V. Ramanathan, R. J. Cicerone, H. B. Singh, and J. T. Kiehl, "Trace Gas Trends and Their Potential Role in Climate Change," Geophys. Res., 90: 5547-5566, 1985.

4. V. Ramanathan et al, "Climate and the Earth's Radiation Budget," Physics Today, 42:(5) p. 22, May 1989,

5. National Academy of Engineering, Energy: Production, Consumption and Consequences, National Academy Press, Washington, D.C., 1990.

6. National Academy of Engineering, Technology and the Environment, National Academy Press, Washington, D.C., 1989.

7. National Research Council, Global Change and Our Common Future, Forum Papers, National Academy Press, Washington, D.C.1989.

8. Greenhouse Warming: Abatement and Adaptation, Workshop Proceedings, Resources for the Future, Washington, D.C., June 14-15, 1988.

9. Energy Policies to Address Global Climate Change, Workshop Proceedings (unpublished), University of California, Davis, September 6-8, 1989.

10. Discussions regarding uncertainties are contained in a number of published reports, including those cited above. For a viewpoint that does not anticipate significant adverse impacts of GHG emissions on climate, see Scientific Perspectives on the Greenhouse Problem, George C. Marshall Institute, Washington, D.C., 1989.

11. J. F. Mitchell, "The Greenhouse Effect and Climate Change," Rev. Geophys. 27:115, February 1989.

12. Data from 1958 to the present are from Keeling's observations at Mauna Loa, Hawaii C. D. Keeling, D. J. Moss, and T. P. Wholf, Measurements of Concentrations of Atmospheric Carbon Dioxide at Mauna Loa Observatory Hawaii, 1958-1986, Final Report for the Carbon Dioxide Information and Analysis Center, Martin Marietta Energy Systems, Inc., Oak Ridge, Tenn. April 1987, and updated by National Oceanic and Atmospheric Administration/Scripps Institution of Oceanography, Boulder, Colo. May 1988). Data for the period 1740 to 1956 are taken from measurements of air trapped in glacial ice sheets (A. Neftel, E. Moor, H. Oeschger, and B. Stauffer, "Evidence from Polar Ice Cores for the Increase in Atmospheric CO_2 in Past Two Centuries", Nature, 315:45-47, 1985.

13. Stephen H. Schneider, "The Changing Climate," Sci. Am., September 1989.

14. Atmospheric Ozone 1985, World Meteorological Organization, Geneva, 1985; see Chapter 3, pp. 56-116.

15. J. P. Riva, Jr., "Fossil Fuels" in Encyclopedia Britannica, vol. 19, 588-612, 1987; "Oil Distribution and Production Potential," Oil Gas J. 86(3):58, 1988); Domestic National Gas Production, CRS Issue Brief, Congressional Research Service, Library of Congress, Washington, D.C., May 2, 1989.

16. World Resources Institute study as referred to by M. Weisskopf in "U.S. Contribution to Greenhouse Effect Rises," The Washington Post, September 16, 1989.

17. Cool Energy: The Renewable Solution to Global Warming, Union of Concerned Scientists, Cambridge, Mass., 1990.

18. "Global Change and Public Policy: A Special Issue," Earth Quest, 3(1), Spring 1989.

19. C. Schneider, "Preventing Climate Change," Issues Sci., and Tech., 5:(4), p. 55, Summer 1989.

3

A FRAMEWORK FOR PLANNING AND IMPLEMENTING ENERGY R&D

In this chapter the status of energy research and development (R&D) in the United States is examined briefly, and insights drawn from past U.S. experience are considered for planning and implementing energy R&D strategies to reduce greenhouse gas (GHG) emissions.[1] Concepts are then outlined to set the framework for the sector-specific analysis and the R&D recommendations presented in Chapter 4.

R&D AT THE U.S. DEPARTMENT OF ENERGY

Funding for energy R&D decreased substantially during the 1980s, from a peak of nearly $5 billion (in constant 1988 dollars) in FY 1979 to about $2.2 billion in FY 1989.[2-4] The cutbacks have been unevenly distributed, as shown in Table 3-1.[5] Research on energy from renewable sources (solar, geothermal, wind) has declined 89 percent since 1979. Nuclear fission, conservation, and fossil energy research have also been cut drastically. A deliberate effort has been made to provide for steady annual increases in the resources devoted to the Basic Energy Sciences program, supporting research and technical analysis.

Cutbacks in many of the applied R&D programs have been justified by intentions to reallocate R&D funding in an "upstream" direction—toward the basic research end of the spectrum. The rationale for this reallocation is that government support is most needed for long-term, high-risk projects that are unlikely to be undertaken by the private sector.[6] Within most programs there has been movement away from advanced development activities and demonstration projects toward exploratory and early-stage applied research. The shift in emphasis from commercialization during the Carter administration to a long-term, high-risk focus of R&D in the Reagan years has caused an erosion in the Department of Energy's (DOE) ability to influence the deployment of new energy technologies in the marketplace. While the basic issues of supply disruptions, rising oil prices, and U.S. energy security continued to persist through the late 1970s into the early 1980s, the energy policies of the Carter and Reagan administrations reflected different perceptions regarding the potential impact of those issues and hence fundamentally different approaches to their resolution. In programs retaining significant amounts of "downstream" activity, industry cofunding has become the norm. An exception to the general emphasis on moving federally funded R&D efforts upstream is nuclear fission research, where part of the effort remains concentrated on downstream activities such as supporting private sector efforts to meet regulatory burdens.

TABLE 3-1 Budget Authority for DOE Civilian Energy R&D Programs[5]

Program	Budget (millions of constant 1988 dollars) FY1979	FY1989	%Change
Solar and other renewables	929	105	-89
Nuclear fission	1,356	293	-78
Electric energy	147	36	-76
Conservation	350	135	-61
Fossil energy	1,035	528	-49
Uranium enrichment	203	115	-43
Biological and environmental research	302	217	-28
Magnetic fusion	327	305	-7
Basic and supporting research	297	438	+47
Energy Information Administration	61	63	+3
Total	5,007	2235	-55

Note: Excluded are activities in general science programs, human genome project, and superconducting supercollider project.

ENERGY R&D OUTSIDE THE DOE

Other Federal Agencies

During the 1980's other federally funded energy R&D programs were scaled back along with those of DOE. From its peak in FY 1981, the Nuclear Regulatory Commission's budget authority for research on nuclear reactor safety and waste disposal declined 64 percent, from $294 million (in 1988 dollars) to $107 million. Since 1979 the U.S. Environmental Protection Agency has cut its funding of energy-related research by 70 percent in real terms, from $175 million to $53 million.[5] Other federal agencies such as Tennessee Valley Authority, Bonneville Power Administration, National Institute for Standards and Technology, the Bureau of Reclamation, U.S. Department of Defense, U.S. Department of Transportation, and the National Aeronautics and Space Administration also engage in energy-related R&D, but they are lesser players.

The Private Sector

Company funds for energy R&D also declined during the past decade but less dramatically than federal efforts. In constant 1988 dollars, company-funded energy R&D fell 30 percent, from $3.46 billion in 1979 to $2.42 billion in 1987, the most recent year for which company data are available.[7] Thus, by 1987, company and federal efforts were of roughly equal magnitude, whereas in 1979 federal funding was 50 percent greater than company funding.

Cutbacks in company-funded R&D, like those of the federal government, have been unevenly distributed across fields of application. Constant dollar company funds for conservation and renewable energy technologies declined 83 percent from $1.6 billion in 1979 to $284 million in 1987. Company funds for nuclear energy R&D fell by 66 percent over the same period, from $262 million to $88 million. Company-funded R&D on fossil fuel technology, however, increased 24 percent from $1.2 billion to $1.5 billion. Thus, in recent years federal and company efforts in conservation and renewables have been of roughly the same scale. Federal resources devoted to nuclear energy have been an order of magnitude larger than company resources, while company spending on fossil fuel technology has been three to five times greater than federal spending. Industrial consortia, other than the Gas Research Institute (GRI) and Electric Power Research Institute (EPRI), have not been major sources of funding for energy R&D even though legislation has reduced the barriers to collaborative projects among companies in the private sector.

GRI and EPRI

Of particular importance within the private sector are the research efforts of two large private consortia, the GRI and the EPRI. GRI's R&D programs (planned and executed under the jurisdiction of the Federal Energy Regulatory Commission) have involved an average expenditure of approximately $150 million per year in recent years. The GRI programs have essentially replaced federal programs in the area of natural gas production, delivery, and end-use conservation.[8] Thus, GRI's and DOE's efforts are complementary.

EPRI (which, unlike GRI, operates outside FERC jurisdiction) funds a variety of R&D programs pertinent to the electric utility industry as well as generic fundamental research. EPRI's annual expenditure for R&D is now around $300 million, of which about $60 million and $35 million are targeted, respectively, at fossil and nuclear power plants. Other major R&D targets include environmental health, safety, and control ($80 million); end-use technologies ($40 million); electricity transmission, distribution, and delivery ($40 million); renewables and energy storage ($15 million).[9]

LESSONS FROM R&D PROGRAMS AND INSTRUMENTS

The committee conducted a limited assessment of the federal energy R&D programs to gather information that could guide the design of future efforts targeted at reducing emissions of GHGs. The assessment excluded DOE's activities under the general science and basic sciences areas. Several structural impediments to effective federal energy R&D management were identified.

A common difficulty is political intervention with the specifics of program design and implementation. In one sense, of course, it would be not just surprising but alarming if elected government officials had no say in government programs. On a deeper level, however, the critique holds that government oversight is unnecessarily compromising the quality of work in energy R&D.

Regional interests often intrude to boost or restrain decisions. For example, the Clinch River Breeder Reactor was funded for many years after a long series of studies, including those of the National Academy of Sciences, found it to be uneconomical. CRBR's longevity was helped significantly by its location. Congress has increasingly viewed the development of energy technologies as public works programs. Sometimes regional interests become so powerful that Congress directs DOE not to consider state and local cost-sharing incentives because this might disadvantage certain states unable to offer them.

Changes in the presidency often lead to major redirection in federal energy programs. In 1977 the Carter administration attempted to dismantle many advanced nuclear projects; four years later the Reagan administration attempted to reinvigorate the same nuclear projects and to terminate the Carter coal and synfuels program. Frequent changes in priorities led to uneconomical, erratic program support as particular issues were of more or less concern to policymakers in the executive or legislative branch. Indeed, the present study is an example of how congressional concern about greenhouse warming and climate change is an attempt to shift energy R&D priorities yet again.

The sensitivity of executive branch programs to changes in the presidency is sometimes countered by congressional attempts at stabilization. An example of this is in the fossil energy program, where 80 percent of projects have been subjected to line-item legislation. Technical management of energy R&D programs could be improved significantly if elected officials in both the executive and legislative branches exercised restraint.

It is axiomatic for good management of large programs that clear and relevant objectives be established before the program is begun and that the program be periodically reviewed relative to its objectives. Many federal energy R&D programs appeared to the committee to lack any clear economic rationale. For example, economic analyses supporting the level of funding or the direction of the fission and magnetic fusion programs were not clearly discernible. In terms of monitoring progress, very little effort appeared to be devoted to understanding whether programs were meeting goals (when goals were specified) or to reallocating funds on the basis of the most promising lines of research. Almost all the federal energy R&D success stories were from programs that had clear objectives.

The lack of clear and defensible objectives and careful monitoring tends to invite politicization, contributes to inertia in R&D programs, and leads to low success rates. As a result, federal energy R&D funds are not being invested as fruitfully as they should be.

Today's structure of federal energy R&D in the DOE has its roots in the nuclear weapons industry, and in many ways the national defense continues to dominate federal energy decisions. This arises in part from DOE's budget. In FY 1989 the total department budget was about $14 billion, of which about $8.8 billion was related to defense and weapons and only about $2 billion was directly related to civilian energy R&D. In addition, the importance of defense issues in national debates has tended to dominate the attention of the top administrator, in turn, of the Atomic Energy Commission, the Energy Research and Development Administration, and the DOE. Currently, with the massive difficulties of cleaning up the wastes in nuclear weapons plants and in starting tritium production, the secretary of energy has little time to focus on overall civilian energy needs. Budget authority for fossil energy, renewables, conservation, and nuclear reactor R&D declined from FY 1979 to FY 1989, while the trend in the late 1980s in nuclear energy R&D was to increase the military component.

In the end the key question concerning the performance of the federal civilian energy R&D program is whether it has succeeded in producing or assisting a significant number of energy technologies to reach technological maturity and market commercialization.

Unfortunately, no comprehensive or well-designed survey of results of federal energy R&D exists, and given the time constraints of this study the committee was unable to undertake such a review. The committee's conclusions, therefore, must necessarily draw on case studies and episodes that may not be representative and on expert opinion that may also be selective. Subject to this reservation, however, DOE's energy R&D programs have shown a low success rate, with few examples of commercialization of technology on a viable long-term basis. However, it should be noted that in the 1980s DOE's criteria and emphasis were increasingly focused on long-term, high-risk R&D with a clear deemphasis on activity close to commercialization.

General lessons from federal energy R&D experience at DOE can be combined with the experiences of programs dedicated to civilian technology development such as at National Advisory Committee on Aeronautics/National Aeronautics and Space Administration[10] and at the U.S. Department of Agriculture[11-12] to provide the following general guidance for the design of GHG reduction R&D strategies:

- To the extent possible, applied R&D programs can and should involve industry participants in establishing objectives and carrying out the research. Competition among firms should be maintained, however, in the commercial development of technologies based on the results of this research.

- Federal research programs function most effectively as a complement to vigorous in-house R&D programs within industry. Especially where such in-house research is lacking, additional funding for extension and other forms of adoption assistance may be critical.

- A decentralized program structure, even though it may be slow to respond to new opportunities or other changes, has important advantages for fragmented industries or for applications that are highly idiosyncratic to varied circumstances.

- A diversified portfolio of publicly sponsored research projects of modest scale is likely to be more effective than a program that concentrates funding among a relatively small number of technologies.

The committee's specific recommendations for improving the management of civilian energy R&D are presented later in this chapter in the section titled, Management of Federal Energy R&D.

TECHNOLOGY DEVELOPMENT AND APPLICATIONS IN OTHER NATIONS

The record of publicly funded R&D programs in many Western European nations is mixed. Many of these programs in aerospace, computers, microelectronics, and (in Great Britain) nuclear energy have suffered from efforts to achieve both national security and commercial objectives within a single program.[13] These programs tended to concentrate funding and technology development efforts on a single "national champion" firm, often constructed from forced mergers among several competitors. Competitive pressure was lacking, and the results frequently were high-cost, noncompetitive technologies.

The scale of government-funded industrial R&D within contemporary Japan has been modest for most of the postwar period. Publicly funded R&D programs in Japan emphasize support for domestic diffusion of scientific and technical knowledge. Cooperative research programs emphasize interfirm diffusion of know-how and incremental improvements of technologies. Cooperation in research is combined with fierce competition among the participating firms in the application of the results of this research.[14] In recent years, however, the willingness of Japanese firms to cooperate in these projects has declined somewhat.

The committee could not review information on the initiation and nurturing of technology R&D programs and applications in developing countries. Important lessons remain to be learned from that experience and applied in future cooperative programs with such countries.

TECHNOLOGY-ADOPTION PROCESS

To achieve successful commercial introduction, R&D programs must frequently be complemented by policies encouraging adoption. Such policies may require the involvement of federal, state, and local governments.

Because the transfer and adoption of new technologies are costly knowledge-intensive activities, public R&D programs are unlikely to develop technologies to technical readiness and then let them "sit on the shelf" until needed. Much modification and improvement occur as technologies are moved into the marketplace. Public funds can be used to subsidize this process through demonstration programs, but care must be taken to avoid the mistakes made in earlier energy demonstration projects. Some federal energy demonstration programs of the 1970s (e.g., Clinch River Breeder Reactor) were too ambitious in pursuing commercial-scale installations in unproven technologies, and they relied too heavily on government management and funding for projects that were heavily oriented toward commercial applications.[15]

ATTAINING LOW-GHG EMISSIONS

Two general concepts govern a move from today's high-GHG economy to a future economy with lower GHG emissions. The first is one in which successful R&D leads to future technological developments favorable to low-GHG fuels and activities that are less costly than comparable high-GHG-emitting fuels and technologies, so the economy naturally makes a transition to a technological base with little potential for climatic impact attributable to GHGs. Under the second concept, governments take stringent measures (such as high carbon taxes or regulations) to move the economy off high-GHG fuels and technologies toward low-GHG ones. As a result, the market price of high-GHG technologies rises relative to those having low GHG emissions. Again, the global economy would tend to shift away from fossil fuels, thereby reducing emissions of GHGs.

The point to emphasize about the two concepts is the difference in the nature of the forces acting on the economy: Under the first concept, reduction of GHGs comes in response to the low costs brought about by successful R&D and technology development; under the second, the impediments or subsidies provided by government policy make fossil fuels uneconomical. Investments in R&D, however, can move the economy more quickly toward low emissions of GHGs and can make the move less painful and less expensive. These concepts are further elaborated in a subsequent discussion of strategy options for energy R&D.

ROLE OF R&D

What R&D (and initial implementation) can produce is information. Alternative energy R&D would reduce uncertainty about the cost, performance, environmental side effects, and other impacts of technologies designed to reduce energy-related GHG emissions. This investment in knowledge serves two purposes. First, it provides a basis for continually redirecting the R&D program toward alternatives that are potentially less costly.

Second, information about new technologies has insurance value and provides a range of options for future deployment, although such deployment will require additional R&D and could take considerable time. Some new technologies may be worth developing, if new knowledge about global climate leads society to place a higher value on reducing GHG emissions. A worthwhile step, therefore, in the interest of quantifying upper or lower bounds on critical cost and performance characteristics of a new low-GHG technology, might be to develop and deploy it on a limited scale, even if its cost exceeds that of the existing energy technologies it would replace. Furthermore, in context, although cost reduction is one stated purpose of an energy technology R&D strategy, there is no guarantee that cost reductions will actually be achieved. Nuclear power is a case in point; its costs increased steadily during the decades following its initial deployment, despite the continued existence of a large federal R&D program.[16]

ENERGY POLICY AND GHGs

In the United States, no fiscal, regulatory, or other incentives are offered to move away from high-GHG to low-GHG fuels and technologies aside from those involving chlorofluorocarbons. Indeed, the economic and energy policies that are in place in most countries, including the United States, tend to be neutral toward, if not in favor of, the continued use of GHG-intensive fuels. Although alternatives to fossil-fuel-based technologies that have lower GHG emissions are available, these are, for the most part, more expensive in the marketplace than high-GHG-emitting fuels and technologies. Given the higher relative cost of low-GHG-emitting technologies in comparison to high-GHG-emitting technologies, and the fact that market prices do not incorporate any cost of future climate change or any benefit from switching to low-GHG fuels, virtually no incentive exists for private firms to invest R&D funds in low-GHG-emitting technologies.

The selection of appropriate policy instruments in the United States for reducing GHG emissions will be strongly influenced by our recent energy experience as well as by new considerations. Notwithstanding industry views to the contrary, a variety of policies in the past spurred development and adoption of technologies. These policies included federally mandated performance standards such as those on corporate average automotive fuel economy and large-appliance energy efficiency; taxation policies, including gasoline taxes and investment tax credits for adoption of conservation and renewables technologies; modification of federal, state, and local regulations, such as building codes; and electric utility regulatory policies that affect private payoffs to adoption.

High taxes on GHG emissions or large tax credits that encourage widespread adoption of low- or zero-GHG emission technologies throughout the economy could be quite costly to the nation. Energy-intensive industries, in particular, would be severely affected. On a national basis, however, these costs could be partially offset by benefits associated with the production of new information arising from experience with the technology in a variety of market segments. Moreover, some segments of the private sector might invest in more R&D because of the larger market created by the adoption incentives. Also, broadly decentralized incentives such as those provided by a carbon tax could identify the technologies that are least costly to develop and implement and would place a cap on the cost of implementing them. Private R&D could be encouraged with tax credits, but this is a relatively inefficient incentive, producing far less than a dollar of incremental R&D per dollar of tax expenditure.[17]

Technology-specific factors will also be important determinants of policy, and experience suggests that future technology demonstration programs for GHG emission reductions should strive for the following characteristics:

- sufficient scale for demonstrating performance and reliability, avoiding premature commitments to commercial-scale projects or dramatic scale-ups;
- substantial industry involvement in project design and management;
- industry cofinancing; and
- concern for and mitigation of unexpected environmental effects.

The issue of global climate change introduces two relatively new considerations into national energy policymaking: how to deal with pervasive and persistent uncertainties and how to include the fact that certain elements of the U.S. strategy may be determined in international negotiations. Because of the uncertainties, preference should be given to energy R&D strategies that are compatible with other national objectives such as economic efficiency and competitiveness, environmental quality, national security, public health and safety, and maintenance of a healthy and flexible economy. Diversity is also a hedge against uncertainty, so U.S. R&D strategy should encompass a broad range of technologies. The strategy should also be flexible in order to accommodate changes in R&D objectives as key uncertainties are resolved. Nevertheless, major irreducible uncertainties will remain and will limit what science can contribute to a national policy response.

At the international level, alternative energy R&D strategies should include policies to assist developing countries in promoting economic growth while minimizing GHG emissions. The global nature of the markets for technologies that produce and consume energy may also create new opportunities for collaborative R&D on alternative energy technologies with other countries.

ROLE OF THE PRIVATE SECTOR

Based on lessons learned from R&D experience relevant to civilian technologies, the private sector will play a vital role in achieving significant reductions in GHG emissions.

Technological innovation is a continuum of activities from basic research through product or process introduction and improvement. The existing state of knowledge of technology at any given time determines the next activity to be undertaken; the market potential determines whether an action should be taken. In general, the federal government's role is important in developing, accessing, and communicating the state of technical knowledge.

Industry should manage the demonstration of technology and its reduction to commercial products and services.

Participation of both the public and private sectors is important throughout all stages of innovation leading to technologies with which GHG emissions can be significantly reduced. Suggested roles for the government and private sector are highlighted in Table 3-2.

TABLE 3-2 Government and Private Sector Roles in Energy R&D and Technology Innovation[a]

Activity	Government Role (federal/state/local)	Private Sector Role (including utilities)
U.S. strategy with respect to global warming	L	A
Energy R&D strategy	P	P
Basic research	L	A
Applied R&D	A	L
Hard-to-capture benefits	L	A
Not-hard-to-capture benefits	A	L
Technology implementation	Market stimulation and intervention	L

[a]L = lead role, P = partnership, and A = advisory role in establishing priorities and providing funding for research, development and demonstration.

MANAGEMENT OF FEDERAL ENERGY R&D

The committee's review of federal energy and other R&D programs summarized earlier suggests ways that the federal government might improve its management of energy R&D. The goal of these management suggestions is to improve the effectiveness of federal energy R&D programs—that is, reduce their costs and increase their benefits.

While there are some notable exceptions, the committee concludes that federal energy R&D programs have often been hampered by conflicting objectives, political interference, inertia in program selection, and preoccupation of top management with defense issues. As a result, the payoff in terms of successful commercialization of civilian energy R&D programs has been modest at best. A clear and more defensible set of project and program management procedures could reduce current temptation for political intervention in program management and project selection. The committee therefore puts forth four recommendations for changes in the management of federal energy R&D that it believes will greatly enhance the social return to federal investments:

- Federal energy R&D programs should establish clear objectives and should systematically reevaluate the individual projects and general direction of R&D in light of these objectives.

- The variety of policy instruments used by the federal government to support energy R&D should be increased. Among the options to be considered are an increase in the peer review of proposals and programs and an increase in the portion of the budget that is open to competitive bidding.

- The management and budget of civilian energy R&D should be insulated from unnecessary political interference and from other programs, particularly those related to defense.

- To improve the management of civilian energy R&D, separation of these programs from both defense and fundamental science now performed by DOE may be advisable.

The sole-source funding of national laboratories should be examined carefully and managed in such a way as to avoid conflicts with R&D programs that are peer reviewed and competitively bid.

Political factors and the defense domination of DOE have tended to reduce the effectiveness of management of civilian energy R&D. While the committee does not recommend removing the energy R&D budget from the normal appropriations process, the selection of projects must be assured on the basis of technical and economic merit rather than political pressure or the existing programs and capabilities of the national laboratories. Stronger DOE evaluation and management procedures could contribute to this goal.

Over two-thirds of the total budget of the DOE is devoted to defense. The high visibility of defense, along with recent difficulties in managing nuclear waste disposal and defense nuclear production facilities, makes it difficult for the top administrators of DOE to give adequate attention to civilian R&D issues. The committee therefore recommends that Congress consider investing DOE's civilian energy functions with accountabilities that are distinct and separate from its defense energy functions. The committee also suggests that Congress consider alternative budget strategies for DOE such as those outlined below.

ALTERNATIVE BUDGET STRATEGIES

Appropriate criteria for budgeting, research, development, and demonstration (RD&D) designed to reduce GHG emissions include stable funding over time and focused attention on the technological merits. These GHG-related criteria must compete with other highly desirable objectives. Among these are substantive policy objectives, such as reducing the deficit and maintaining economic competitiveness, and procedural objectives, including facilitating choices among national priorities and making these choices transparent. There is also an implied objective in maintaining the integrity of the process by revealing all costs on an equal plane.

All objectives could be met, given necessary political agreement, by a lump-sum annual appropriation to a lead agency, which would then divide this budget authority among the participating units. The funds would be available; they would be provided in a public and, therefore, accountable manner; and the program would be fiscally responsible.

There are other ways of funding programs outside the appropriations process. Private research, for instance, could be funded by tax credits. In this way its funding is automatic, and this tax expenditure is not formally counted toward the deficit, though, of course, it does reduce revenues. Trust funds could be established based on earmarked taxes. The fate of the highway trust fund, however, warns against the premature view that such funding is guaranteed.

A multiyear or "no-year" appropriation could be sought. By taking funding out of the appropriations process, instability in funding might be avoided. Yet that stability could be obtained, given the deficit, only at the expense of other programs. In any event, nothing can stop Congress from reconsidering any time it chooses. It is the public political support behind the programs that matters.

There is no need to rely on a single form of funding. If, say, tax credits were deemed superior for private industry, appropriations could still be used by in-house government and noncorporate researchers. However, tax credits are difficult to target appropriately.

The committee's preference is for a maximally visible fund through the annual appropriations process.

LEVERAGING FEDERAL INVESTMENTS GLOBALLY

The global character of the GHG issue imposes a special requirement on R&D. Both the advancement of science and the development of alternate "solutions" require an international context. The foreseeable R&D costs to make progress in these two areas will be high; hence, it would be desirable to share these costs as broadly as possible and to seek priority solutions that offer the greatest promise for GHG emission management should they be needed.

International cooperation has been an established tradition in the natural sciences. Research on climate has involved major international experiments; these are ongoing programs that will yield important results about climate change. Analogous programs aimed at technology development to respond to GHG management have been discussed in ad hoc forms. Only recently, through the Intergovernmental Panel on Climate Change (IPCC), have talks begun on an international technological response. The IPCC deliberations should generate at least a framework for international RD&D; however, it is unlikely that the IPCC will institutionalize such a program.

Parallel to the IPCC activity, there are ad hoc industry discussions taking place. These are building on informal cooperative relationships developed in the electric utility industry and the petroleum and gas sector. Existing arrangements facilitate a continuing development of non-fossil-fuel electrical generation opportunities as well as transportation options. Examples of such arrangements include the United States-United Kingdom-France-Japan cooperation in developing safe nuclear power and the U.S.-Dutch coal generation demonstrations. A different example is the independent Japan-European R&D leading to improved fuel-efficient internal combustion engines. These focus on the applications and needs of the developed world but not on the needs of developing counties.

Industry discussions in the United States concerning R&D on energy use and emissions are also under way but are outside the context of climate change at the present time. Most recently, 14 of the largest oil companies have joined the three domestic automobile manufacturers in a collaborative R&D effort on alternative transportation fuels for the United States. The results of this collaboration are bound to have international significance.

Analysis of the future energy options for the world indicate that, although considerable gains on GHG emissions are achievable in the developed countries, the greatest leverage on future change is in developing countries. The latter is set in the context of projected increases in population, progress toward economic parity, and greater industrialization.

A major opportunity is at hand for RD&D in cooperation with the developed countries to seek options for energy supply in the developing world. This must take advantage of resources such as forest management; simple, small-scale, efficient electrical generation; efficient public transportation; high end-use efficiency; and emission control schemes that will permit continued exploitation of world coal resources. At present the options are available for technical means to arrest GHG emissions while providing for energy needs. However, the means for demonstrating the feasibility and reliability of alternatives has not been provided for. Demonstration of technology is an expensive and long-term commitment. To enable such demonstration, a cooperative program between government and the private sector would be an important element of U.S. energy strategy.

International cooperation in energy RD&D can be encouraged through governmental arrangements or by ad hoc agreements with energy producers. Government commitments have been facilitated by formal programs of the United Nations, by informal arrangements using national laboratories, or by ad hoc organizations such as the International Institute for Applied Systems Analysis or the Center for European Nuclear Research. In the energy sector, commitments have been made through such institutions as the National Academy of Sciences, the GRI, and the EPRI, linked with sister organizations like the foreign academies of sciences and the foreign electric and gas research laboratories. The latter organizations are well suited for and experienced in the development and management of demonstration programs. The strategy choices for a U.S. RD&D effort should take advantage of this experience.

STRATEGY OPTIONS

Two energy R&D strategies together with market intervention policies and actions are available to the United States for achieving reductions in GHG emissions:

- Focused R&D Strategy. Pursue energy R&D that is aimed at reducing GHG emissions and that would make sense for other reasons even in the absence of concerns about global climate change.

- Insurance Strategy. Pursue energy R&D that would be viable only in the presence of concerns about global climate change.

Both strategies follow the conventional R&D paradigm of reducing uncertainties about the cost and performance of a technology by producing new knowledge. They address, in general terms, roughly the same set of technologies (encompassing the entire fuel cycle from supply through utilization), and span the full range of activities from fundamental research to technology adoption, but they differ in purpose, cost, and policy instruments.

The economic rationale for the Focused R&D Strategy is based on the inability of private firms to capture the benefits of basic research and the fact that the price of fossil fuels is less than the full social cost associated with their use. The rationale for the Insurance Strategy is that additional energy R&D is warranted by the conditions that prevention of climate change may assume a high priority in the future and that new technologies would be needed to reduce GHG emissions. The Insurance Strategy is incremental to the Focused R&D Strategy. The fundamental difference between them is the difference in the magnitude, timing, and costs of actions that can be justified on non-GHG grounds and those that need a GHG justification.

The federal R&D program under the Insurance Strategy will be considerably more costly to the government (involving multibillion dollar increments over the Focused R&D Strategy), and a greater fraction of the government's R&D would be directed toward reducing the uncertainties associated with the technology-adoption phase. Through the Insurance Strategy the nation would, over time, invest in the development and demonstration of a variety of "backstop" technologies for their "insurance" or option value.

For example, before a new type of nuclear reactor becomes viable, it might have to be sited, licensed, and successfully operated for a number of years in order to convincingly demonstrate that dealing with safety and environmental concerns would not substantially increase the real cost of the technology, as occurred in the case of the light water reactor.[16] Similarly, the government may need to underwrite the costs of demonstrating the economies of mass production of advanced batteries or the biological sustainability and environmental acceptability of large-scale biomass plantations in various regions of the country. Resolving uncertainties associated with the "infrastructure" for these new technologies may require full-scale testing in certain market niches (e.g., fleets of electric or biomass-fueled delivery vehicles or rental cars) or in government programs (e.g., silviculture for erosion control) where the existence of other benefits may reduce the cost of implementation.

The Insurance Strategy need not be directed exclusively on alternative energy technologies aimed at the domestic market. Technologies suitable for use in other countries such as India and China could have leverage for affecting worldwide GHG emissions. Furthermore, the strategy may include the development and demonstration of CO_2 sequestering technologies that would be of little interest in the absence of concern about GHG-induced climate change.

It is not sufficient to define energy R&D priorities in isolation from the marketplace for which the products of the R&D are intended. Prevailing market forces must be considered and government actions may be required to achieve specific national objectives. In the past, particularly at times of crisis, the government has used intervention mechanisms such as taxes, tax credits, energy efficiency standards, loan guarantees, subsidies, federal procurements, and liability limitations to influence the supply and demand of fuels and energy resources. In the event that the nation makes a commitment to reduce emissions of GHGs significantly, such actions ought to be considered again as a supplement to the Focused R&D and Insurance strategies. This would stimulate energy R&D in the private sector and the adoption of GHG-reducing technologies in the marketplace.

In the near term (i.e., from the year 1990 to 2000), such actions could spur the adoption of GHG-reducing technologies that already exist and that can be shown to be economically viable for reasons other than low-GHG emissions but that are not currently being used. For example, policies may be needed to influence the regulatory environment at the state and local levels and facilitate widespread adoption. Tax and regulatory policies may be warranted, because they yield net benefits consistent with reliability of energy supply and other national goals such as security (e.g., an oil import tariff or automotive fuel economy standards) or economic efficiency (e.g., promote investments in energy-conserving equipment or buildings).

Market intervention could also be formulated to shift the entire burden of applied energy RD&D the private sector. For example, a carbon tax could stimulate development of alternative technologies by making fossil-fueled vehicles more costly to operate. It could elicit a diverse R&D response from the private sector and facilitate an efficient transition (e.g., encouraging the use of methanol made from natural gas while biomass plantations were becoming established). A carbon tax could also send clearer signals to the market about the relative costs of electric and biofueled personal transportation systems, as electricity prices began to reflect the costs of generating technologies having low- or zero-GHG emissions. Such actions could enable the private sector to capture the benefits that the nation may attach to reducing GHG emissions. They would, however, still leave the government with its traditional role of performing basic generic research because its benefits cannot be appropriated. A pure market intervention strategy would not change what needs to be done by way of energy R&D but would shift to the private sector the responsibility for its planning and execution.

Chapter 4 draws on the framework presented in this chapter, and defines alternative energy R&D programs and actions to achieve reductions in GHG emissions. The committee's recommendations are not governed by explicit objectives to achieve specific levels of reductions in U.S. GHG emissions over different time horizons. However, the technology-adoption actions identified in the various market sectors relate to existing technologies that can be shown to be reasonably cost-effective (i.e., economically viable aside from their GHG emissions reduction value) and to technologies in R&D once the uncertainties regarding their cost and performance have been reduced to acceptable limits.

NOTES AND REFERENCES

1. For a detailed treatment of federal energy R&D, see R. G. Hewlett and B. J. Dierenfield, The Federal Role and Activities in Energy Research and Development, 1946-1980: An Historical Summary, Oak Ridge National Laboratory, Oak Ridge, Tenn., 1983.

2. Fiscal Year 1990 Budget Highlights, U.S. Department of Energy, Washington, D.C., January 1989, p. 4.

3. Federal R&D Funding by Budget Function: Fiscal Years 1989-1990. NSF 89-806: National Science Foundation, Washington, D.C., April 1989.

4. The Advanced Nuclear Systems ($38 million) and the Space and Defense Power Systems programs ($66 million) are directed entirely at NASA and military applications. An unspecified fraction of the expenditures on the Test Facilities ($138 million) program is also directed toward noncivilian applications. See DOE's Fiscal Year 1990 Budget Highlights, pp. 15-16.

5. Federal R&D Funding by Budget Function, National Science Foundation, Washington, D.C., various years.

6. Private sector underinvestment in R&D occurs not because projects are long term and high risk but because marginal social returns exceed marginal private returns. Such circumstances arise because (1) the marginal returns to R&D cannot be fully appropriated by the innovator (e.g., there are spillovers to competitors) or (2) the products or services on which R&D is focused are unpriced or inappropriately priced in the market (e.g., market prices fail to reflect environmental damages or premiums for national security). To the extent that the results of long-term, high-risk projects are less appropriable than the results of short-term "downstream" projects, such projects are prone to underinvestment by the private sector and may warrant government support.

7. Data on company-funded R&D were supplied by the Science Resources Section of the National Science Foundation.

8. Historical Review of Gas Research Institute Research and Development, Gas Research Institute, Chicago, Ill., May 1987.

9. Research and Development Program 1989-1991, Electric Power Research Institute, Palo Alto, Calif., January 1989.

10. The discussion of NASA's research programs draws on D. C. Mowery's paper presented at the NAS workshop on the returns to federally funded R&D, November 1985; and D. C. Mowery and N. Rosenberg, Technology and the Pursuit of Economic Growth, Cambridge University Press, New York, 1989, Chapter 7.

11. R. E. Evenson, "Agriculture," in R. R. Nelson (ed.), Government and Technical Progress, Pergamon Press, New York, 1983, provides the basis for this paragraph.

12. "Overall, the experiment stations have generally moved their work into areas where they have a comparative advantage vis-a-vis the private sector. In direct competition with market-oriented private firms, the public sector does poorly and generally does not invest heavily in research of that type. It tends to be pressed into work of a testing and certifying nature, designed to help farmers make choices among suppliers of inputs. In recent years it has played a major role in facilitating adjustment to regulations both in the chemical inputs fields and in food technology" (Evenson, op. cit., p. 275).

13. R. R. Nelson, High-Technology Policies: A Five-Nation Comparison, American Enterprise Institute, Washington, D.C., 1984. The French nuclear program, which shares many of the undesirable features of the British nuclear program and both the British and French programs in computers and aircraft, appears to have been relatively successful in producing reactors for extensive domestic use. This success was aided by the existence of a state-owned monopolistic domestic customer for the French reactors, which facilitated design standardization and reduced regulatory obstacles to adoption. Framatome, the major French producer of reactors (also state owned), nevertheless does not appear to be highly successful as an exporter in world markets.

14. See D. Okimoto, "The Japanese Challenge in High Technology," in R. Landau and N. Rosenberg (eds.), The Positive Sum Strategy, National Academy Press, Washington, D.C., 1986, and Mowery and Rosenberg, op. cit., Chapter 8.

15. J. F. Ahearne, Why Federal Research and Development Fails, Discussion Paper EM 88-02, Resources for the Future, Washington, D.C., July 1988.

16. Nuclear Power in an Age of Uncertainty, Report OTA-E-216, U.S. Congress, Office of Technology Assessment, Washington D. C., February 1984.

17. Tax Policy and Administration: The Research Tax Credit Has Stimulated Some Additional Research Spending, U.S. General Accounting Office, Washington, D.C., September, 1989.

4

POTENTIAL FOR REDUCING EMISSIONS OF GREENHOUSE GASES

The potential for reducing greenhouse gas (GHG) emissions associated with the production and use of energy are analyzed in this chapter from the standpoint of four market sectors: electric power, transportation, residential/commercial buildings, and industry. A key part of this analysis is the identification of alternative energy technology options to meet specific service demands in the respective market sectors such that the technology application is accompanied by significant reductions in the quantity of GHGs emitted per unit of service provided, compared to current practice. In this sense the electric power sector is concerned with the generation, transmission, and distribution of electricity to all users; the industry sector is concerned with the manufacture of all products, including fuels; the transportation sector is concerned with technologies to move people and goods; and the residential/commercial sector is concerned with the design of buildings for all sectors as well as with the provision of services within the building envelope (space conditioning, lighting, refrigeration, etc.) and the efficient utilization of the pertinent technologies.

The time frames of relevance to this study include the near-term period through the year 2000 and the long-term one that goes to the year 2050 and beyond. Each sector analysis encompasses actions for achieving commercial adoption (implementation) in the near term of promising energy technologies for which essentially no R&D is required and simultaneously identifying R&D needs, priorities, and implementation strategies with the potential for high payoff over the long-term horizon. Recommendations are formulated within the confines of activities and services relevant to each sector and are expressed in terms of selective changes to the current federal energy R&D agenda. However, no specific GHG emission reduction objectives versus time have been postulated for the technology-adoption actions that are recommended.

Most of the substance of this report is based on the experience and expertise of the members of the committee and panels and on information obtained from various sources (see Acknowledgments). When appropriate, the committee and panels made use of prior studies on energy technologies for reducing GHG emissions, (see Notes and References and Bibliography at the end of this chapter). In parallel with the current study, the U.S. Department of Energy's (DOE) national laboratories were preparing white papers on energy efficiency, renewable energy, global climate change, and technology transfer for consideration in the national energy strategy. While these papers were not all available to the committee and the panels during their deliberations, they are cited in the Bibliography.

ELECTRIC POWER

Energy Use and GHG Emissions

The electric power sector has the potential to produce and deliver electricity essentially free of GHG emissions, primarily CO_2. Currently, however, electricity is generated worldwide predominantly from fossil fuels, with coal being the dominant fuel choice. Non-CO_2-emitting electricity-generating technologies based on nuclear fission reactors, renewable sources, and geothermal energy are commercially available and technically could supply the world's energy needs. Because of unfavorable economics as well as environmental, health, and safety concerns, however, it is by no means clear that these technologies could be deployed on the scale required without substantial research, development, and demonstration (RD&D) and costs. The transition from coal to these non-CO_2-emitting sources will involve major changes in the operating and performance criteria applied to the selection and deployment of generating technologies in energy markets served by the electric utilities and in the economy as a whole.

The electric power equipment that is in place in the United States and that supplied power in 1988 is shown in Table 4-1. The major source of CO_2 (over 85 percent) is from coal because of its high carbon content per British thermal unit and its use in base-load operations. Electricity production results in the annual emission of over 450 million metric tons of carbon (MTC) annually as CO_2 and only a relatively small amount of other GHGs. Hence, this chapter is primarily concerned with measures and R&D strategies to reduce CO_2 emissions in the electric power sector.

Major Targets for Attention

A CO_2-free electricity supply system is feasible based on the technology available today. Such technology is currently uneconomical or unacceptable to large segments of our society.

The social and environmental impacts of these technologies and the related political and regulatory factors are as critical as technical options if CO_2 emissions are to be significantly reduced. Unless these factors are appropriately addressed, the money allocated to "solving" technical problems will be wasted.

For a significant reduction in GHG emissions, demand for electricity must be reduced and an effective generating strategy implemented. The alternative pathways are shown in Figure 4-1.

TABLE 4-1 1988 U.S. Electric Power Generation[1]

Fuel	Carbon Emissions million metric tons	Generation 10^9kwh	Percent of Generation
Coal (758 x 10^6 short tons)	398 (85%)	1,538	57.0
Oil (248 x 10^6 barrels)	31 (7%)	149	5.5
Gas (2,634 billion ft^3)	39 (8%)	252	9.3
Nuclear		526	19.5
Hydro		223	8.3
Renewables	Negligible	12	0.4
Total	468	2,700	100.0

Note: The CO_2 emissions per unit of fuel consumed in the generation of electricity are as indicated in Section 2, Problem Description.

Such a strategy would aim to increase the relative percentage of electricity produced by non-CO_2-emitting generation options. In the near term use of energy from existing alternative resources would be maximized. In the longer term the strategy would deploy alternative generation options for new plants and retrofits, coupled with an environmental dispatch strategy to use all generating capacity in the most environmentally effective manner. Approaches that could help reduce CO_2 emissions are listed below for the near and long term.

Significant reductions of CO_2 emission from current levels may be possible in the near term if the following actions are implemented between now and the year 2000:

- Increase end-use efficiency in all end-use sectors by aggressive R&D and demand-side management programs.

- Increase nuclear power plant availability from the current 64 percent to the highest international practice (75 percent to 85 percent).

- Resolve the controversies that are currently delaying operation of those nuclear generation facilities that are complete or nearing completion.

- Increase the efficiency of existing fossil fuel units by improved operation and maintenance.

- Cofire natural gas with coal and substitute natural gas for coal to the extent it is available.

- Encourage the installation of cogeneration units to increase the overall efficiency of combined heat and electricity production.

- Improve existing transmission and distribution facilities (with both alternating current and direct current additions) to increase the efficiency of the network and permit greater gains from a larger and more efficient environmental dispatch.

- Adopt environmental dispatch approach of power when feasible.

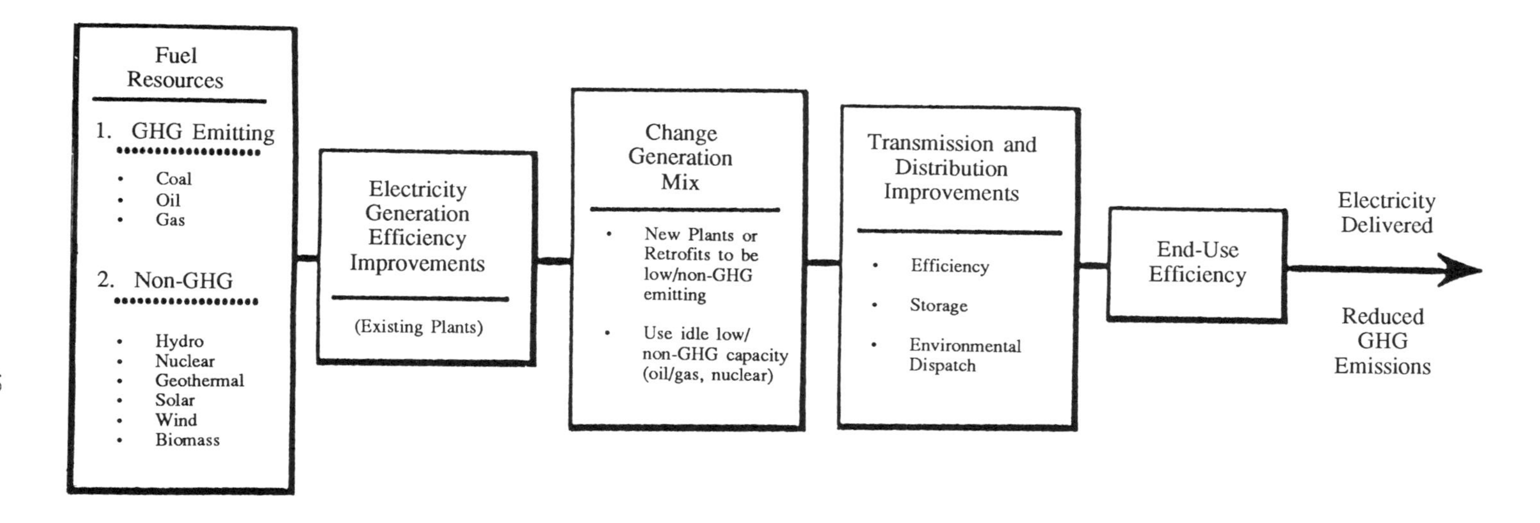

FIGURE 4-1 Reduction strategies for GHGs emitted in the electric power sector.

Aggressive development of near-commercial technologies between now and the year 2025 will assure that non-CO_2-emitting facilities will be available for installation in the post-2025 period. Their deployment could eliminate emissions of GHGs from the electric power sector by 2050. The long-term actions are as follows:

- Sustain aggressive end-use efficiency improvements in all sectors to offset growth in the demand for electricity.

- Develop and commercialize one or more advanced nuclear power reactors that are acceptable to the global market.

- Develop and commercialize renewables to the extent feasible.

- Improve the efficiency of existing hydro installations by replacing inefficient facilities with modern high-efficiency equipment.

- Retire CO_2-emitting power generation equipment as rapidly as non-or low-emitting alternatives can be installed.

- Promote regulations to encourage adoption of non- or low-CO_2-emitting generation facilities.

Availability of Technology to Reduce GHG Emissions

The technology evaluations in this chapter are subdivided into the primary energy resources—fossil fuels, nuclear energy, and renewable/unconventional energy. In addition, transmission, distribution, and storage technologies are addressed.

Fossil Fuels

The new fossil fuel technologies under development that will use coal are a set of fluidized bed technologies (atmospheric fluidized bed, pressurized fluidized bed,) and the integrated coal gasification, gas turbine, combined cycle system (IGCC).[2] The overall thermal efficiencies of these new coal-based technologies are equal to or better than the existing pulverized coal plants and can have reduced emissions of environmental pollutants (SO_x, NO_x, and particulates).

Of the new coal-based technologies, the IGCC system has the highest efficiency and the lowest emission of environmental pollutants. Reduction in CO_2 is directly related to thermal efficiency gains; thus, the IGCC has the potential for being the preferred coal-based technology considering all environmental emissions, including CO_2.

To reduce CO_2 from natural gas and coal combustion, the development of improved combined cycles and other advanced gas-turbine-based power technologies is essential.

Clean coal technology is becoming more important and will have an impact on all new coal-fired power plants and many existing ones. Some of the clean coal technologies will achieve major reductions in SO_x emissions at the expense of power plant efficiencies and a consequent increase of CO_2 emissions. Research in these clean coal technologies should favor avenues that do not carry penalties in the form of increased CO_2 emissions.

Environmental research and regulations should therefore include GHG emissions as a criterion in evaluating new approaches to coal combustion.

Capture and Disposal of CO_2

Although capture of CO_2 emissions from combustion gases can be achieved by conventional technologies—with assessment of energy penalties in the 15 to 30 percent range [3,4]—this is only a small part of the problem. Disposal of CO_2 is likely to be much more costly and may ultimately impose a prohibiting energy penalty. A number of options have been proposed, such as disposal in the deep oceans or in abandoned gas wells. Their feasibility needs to be assessed.

Nuclear Energy

Nuclear power is an important alternative to energy from fossil fuels and a potentially important component in a low-CO_2 emission strategy. It can be an efficient source of energy capable of generating electricity and/or process heat. Light water reactor (LWR) technology is highly developed and mature in comparison to other renewable or nuclear alternatives, but continued deployment of nuclear technology in the United States is fraught with hurdles that impede nuclear power as a major option for reducing CO_2 emissions.

Two technical approaches to these problems are often suggested: one based on evolutionary improvements to existing LWR designs and the other based on new designs, almost revolutionary in approach. Proponents of an evolutionary strategy believe that the option of a major shift away from conventional LWR technology is unrealistic and illusory. They argue that it is wiser to draw on the great store of LWR experience in order to move incrementally toward an improved LWR system than to forgo this experience in favor of an unproven concept(s). Moreover, they point out the great difficulty of changing the technological course of an industry that for over three decades has been so strongly oriented toward LWR systems.

On the other hand, the evolutionary approach by its very nature may be insufficient to address the fundamental problems that have arisen with nuclear power in the United States. A radical technological shift need not entail a completely new start, for a good deal of the existing LWR technology base is likely to be transferrable irrespective of the direction of the shift. In the case of LWR, liquid metal reactor (LMR), and modular high-temperature gas reactor (MHTGR) systems, relevant experience exists both in the United States and overseas. Finally, disaffection toward conventional LWR technology in the electric utility industry and among the general public may be so strong, and the managerial and regulatory difficulties of the existing industry so great, that only a radical technological change could help restore the nuclear option.

Basic changes in the assumptions and policies of industry and government will be required to stimulate a more vigorous technological response to the current problems facing nuclear power.

Current Advanced Reactor Development. Recent technology advances achieved by various programs conducted or sponsored by DOE, the U.S. private sector, the European community, Japan, and the U.S.S.R. are leading to the development of new generations of reactors. Most of the advanced reactors fall into one of six types:

- evolutionary large light water reactors (LWRs),
- advanced passive medium-sized LWRs,
- conceptually new LWRs,
- heavy water reactors (HWRs),
- modular high-temperature gas reactors (MHTGRs), and
- liquid metal reactors (LMRs).

Fusion Technology. Of the several approaches investigated since controlled thermonuclear research started, two stand out as the most promising: the inertial laser fusion reactor and the Tokomak magnetic fusion. The bulk of the effort on laser fusion is sponsored by the U.S. Department of Defense, while the Tokomak is mostly funded by DOE. In both the United States (Tokomak Fusion Test Reactor) and the European community (Joint European Torus), fusion reactors are within a factor of 2 to 3 from the break-even point. This represents a millionfold improvement over a span of 20 years. However, even if a self-sustained experimental reactor is demonstrated, fusion reactors face serious engineering problems related to superconducting coil designs, materials issues of radiation damage, and technology of energy extraction. Conservatively, magnetic fusion reactors will not be a significant component of the U.S. electricity generation mix before the year 2050.

Institutional and Technological Constraints. Among the major institutional deterrents to large-scale introduction of nuclear power, foremost are high cost, poor public acceptance for reasons of safety, and poor fit with utility systems. If nuclear power is to be developed on a large scale, additional institutional changes may be required to make the nuclear energy system diversion resistant, including internationalization of certain parts of the nuclear fuel cycle. Finding institutional arrangements that will both make the nuclear system acceptably diversion resistant and politically acceptable will be challenging and will probably require major international cooperation.

Renewables

Renewable technologies can be classified as renewables with inherent storage capacity (hydro, biomass, geothermal, ocean thermal) and intermittent renewables (wind, solar thermal, photovoltaics). Aside from hydro, renewables in 1988 accounted for 0.4 percent of electricity generation in the United States. Renewables, however, offer the potential for significant exploitation in the future, with environmental benefits and promising economics.

Hydro. Some 64 gigawatts (GW) of conventional hydro and 17 GW of pumped storage capacity have been developed in the United States. The latent potential has been estimated as 75 GW conventional and 15 GW pumped storage, but the environmental costs of this development could be severe.[5] Plant efficiencies could be improved with new variable-speed, constant-frequency generators. The national potential for such upgrades needs to be determined.

Biomass. Burning biomass grown renewably makes no net contribution to atmospheric CO_2. Most of the present 8 GW of installed biomass generating capacity in the United States is based on the steam Rankine cycle and is concentrated in the pulp and paper industry, where the fuel used is low-cost wood wastes.

Some pressurized airblown gasifiers closely coupled to various steam-injected aeroderivative gas turbine cycles appear to be well suited to biomass applications. The coal gasifier/combined cycle technology demonstrated at Cool Water, California,[5-8] may have lower unit capital costs with biomass versions. This may be because biomass generally contains negligible sulfur, the removal of which is costly for coal systems. If these potential advantages could be realized, electricity produced from biomass could be competitive with electricity from conventional coal steam-electric plants in situations where sustainable management of the resource is cost-effective.[6]

Geothermal. The U.S. geothermal industry is presently producing 3 GW of baseload electricity. The U.S. Geological Survey has estimated the total U.S. hydrothermal resource usable for power generation to be 2,400 quads, located primarily in the western states, Alaska, and Hawaii. The U.S. geopressured resource—the energy in overpressured reservoirs of hot water that contain dissolved methane—is estimated to be 180,000 quads. If heat mining of deep hot dry rocks can be developed, between 10^5 and 10^6 quads of energy might be available.

Ocean Thermal Energy Conversion (OTEC). Potential commercial opportunities of OTEC are primarily outside the United States. Basic research on biofouling and corrosion in marine environments, exploratory research on low-temperature differential thermal cycles, and systems studies relating OTEC to other non-GHG-emitting technologies may prove valuable. The committee did not evaluate OTEC programs and has no recommendations on what R&D is most appropriate.

Wind. There is 1.5 GW of installed wind capacity in California. There has been a fourfold reduction in the cost of wind power from the best new wind farms since 1981. Further cost reductions could arise with new technological advances in composite materials, manufacturing processes, and "smart" controls.

The wind resource is less dependent on latitude than other solar sources. The accessible U.S. resource has been estimated to be 1,000 times the electricity currently produced by wind.[7,8]

Solar Thermal. For high solar insolation areas, solar thermal-electric technology is promising. The parabolic trough is the most mature solar thermal-electric technology with 200 MW of capacity currently operating on a utility grid in California in the hybrid mode (with natural gas backup). Some 80 MW of additional capacity is under construction in the United States and 320 MW is planned. Variants on solar thermal technology include the parabolic dish/dish-mounted engine generator and the heliostat/central receiver system.

Photovoltaics (PV). The price of photovoltaic modules has fallen from about $120 per peak watt (in nominal dollars) in the early 1970s to the range of $4 to $5 per peak watt today. Attractive features of PV technology are that no cooling water is needed for flat plate collector systems and prospective economics are favorable at small scale.

One approach to reducing costs further involves the use of thin films, which promise very low unit capital costs because of the tiny amounts of active materials involved and the suitability of the technology to mass production techniques. Amorphous silicon, copper indium diselenide, and cadmium telluride are the leading competing thin-film technologies.[9,10] Alternatively, high-efficiency crystalline cells could be used in tracking and concentrating collectors.

R&D Needs and Priorities

Following are broad guidelines for developing an effective set of options for generating electricity with technologies that will significantly reduce emissions of GHGs:

- The most important and immediately effective option is increasing energy productivity that is beneficial in addition to its potential for GHG reduction.

- A number of nonfossil energy options are possible for GHG emission reduction in power generation. All technically feasible and environmentally acceptable options should be pursued. A multiple option strategy in energy policy is critical for its success.

Increased and consistent R&D funding is required to develop and deploy the most promising low- or non-CO_2-producing electric generation technologies. While increased funding is necessary, it is not sufficient.

New mechanisms are needed involving government, industry, and the electric utilities. Time constraints for this study did not allow the development of such mechanisms. In general, however, R&D efforts should be concentrated to the extent possible in the private sector either by direct funding of private performers or indirectly by policies that increase private returns for publicly needed R&D.

- Early action in developing and implementing electricity supply strategies will help minimize later difficulties because the time from technology conception to its widespread adoption is measured in decades.

- The trend toward a more electric future, as well as the fact that most nonfossil energy options produce electricity, indicates the need for and benefit of research on future electric systems—storage, interactive load control, increased efficiency, and regional interconnections.

- International cooperation is needed to develop energy strategies and to promote consensus for a set of global energy technologies. U.S. leadership and early initiative will have a demonstrative effect elsewhere and hence a multiplier effect worldwide.

A major point to be considered is that accelerated market adoption of technologies requires a coordinated institutional approach involving government, manufacturers, utilities, and regulators. In addition, economic evaluation concepts which internalize environmental costs are needed if minimum CO_2-producing technologies are to become the "economic" choice. Although both the federal and private sectors may have effective programs, they are seldom combined in ways that will produce the maximum technology development per total dollar expended. Considering the magnitude of the technology development needed to bring forth commercially available non- or low-CO_2-emitting technology for electric power production, a vastly superior federal/private collaboration in RD&D is needed.

Review of DOE Fossil Energy R&D Program

The DOE Fossil Energy R&D Program has had as its main purpose the continued use of fossil fuels as a fundamental component of our domestic energy use. The program has two parts: clean coal technology funded in FY 1990 at $554 million and fossil energy R&D funded in FY 1990 at $417 million.[11] The main thrusts of the clean coal technology efforts, which are cofunded by other participants, are the reduction of acid gas emissions with emphasis on retrofit technologies and projects that will be ready for application by 2005. The fossil energy R&D efforts include longer-range projects and research on improved and unconventional hydrocarbon recovery projects. The effect on carbon dioxide emissions was not a criterion during the selection of the present projects.

The IGCC and fuel cell could provide CO_2 emission reductions of 20 percent with currently available technology; and much greater reductions are possible using advanced gasifiers, and high efficiency advanced gas turbines, or advanced fuel cells.

Most of the flue gas scrubbing technologies increase the energy consumption and thus the CO_2 emissions per kilowatt hour. Coal gasification projects also produce extra CO_2 during the gasification process. If the gas is used in conventional furnaces, a net increase in overall CO_2 production will occur. The burden of any overall CO_2 reduction will fall on the generating system, IGCC with or without fuel cells. Government funding of combined-cycle systems has been and continues to be very limited.

Fossil Energy R&D Needs

To obtain maximum efficiencies from gas turbine systems, accelerated R&D is needed on high-temperature-operation; advanced blade cooling technologies; intercooling, reheat, and regenerative cycles including steam-injected cycles; and chemically recuperated cycles. R&D of improved gasifiers, both for coal and biomass, is required for high-efficiency IGCC systems.

A long-term option that should be studied is the capture of CO_2 from combustion processes. The United States should pursue basic generic research on the subject and keep abreast of developments worldwide, including in Japan.

Review of DOE Nuclear Energy R&D Program

The current DOE civilian nuclear fission R&D program has a threefold objective:[12,13]

- to develop the advanced LWR in cooperation with industry to meet near-term needs,

- to develop advanced reactors (MHTGR, LMR) for the longer term, and

- to eliminate overly restrictive regulatory barriers to the use of nuclear power.

The current DOE R&D funding is shown in Table 4-2. The main funded activities are summarized below:

- Evolutionary LWRs

DOE funding is limited to safety analysis and support for the Nuclear Regulatory Commission's submittal of the Safety Analysis Report. Support is given to General Electric's (G.E.) advanced boiling water reactor (ABWR) and to Combustion Engineering's System 80+.

- Advanced Passive Medium-Sized LWRs

DOE funding is divided here between the G.E. small boiling water reactor (SBWR) and the Westinghouse AP600. The program is directed at both separate safety tests and initiation of detailed design of these 600 MW reactors.

- Modular High-Temperature Gas Reactor (MHTGR)

The major part of DOE funding is directed to support the completion by General Atomics of their preliminary design efforts, licensing interactions with the Nuclear Regulatory

Commission, technology development efforts to demonstrate quality of fuel elements, and participation in international collaborative agreements. The remaining allocation supports the Oak Ridge National Laboratory's work in fuel and material technology developments.

- Liquid Metal Reactor (LMR)

The DOE advanced LMR program consists of two separate but related parts: the Integral Fast Reactor program at Argonne National Laboratory and the small advanced liquid metal reactor, PRISM, under conceptual design by G.E. DOE funding for Argonne focuses on technology demonstration of metal fuel, demonstration of processing technology, and fuel recycle capability. DOE support for G.E. is primarily in the preliminary design efforts for PRISM. Additionally, DOE funds Oak Ridge's efforts, in collaboration with Japan in the oxide fuel reprocessing development programs.

The large proportion of DOE advanced reactor expenditures on facilities support raises questions and indicates an imbalanced program.

TABLE 4-2 DOE Advanced Reactor and Magnetic Fusion Appropriations[11,14]

Activity	Appropriation (Millions of dollars) FY 1989	FY 1990
Large LWR pants	6.5	3.4
Mid-sized LWR plants	8.5	15.0
Other LWRs	2.5	0.0
MHTGR	19.8	22.5
LMR/IFR	54.2	35.1
Total advanced reactors	91.5	76.0
Facilities support	139.6	169.6
Magnetic fusion	344.6	320.3

Nuclear Fission R&D Needs

A separate study by the National Academy of Sciences (NAS) on future U.S. nuclear power development is currently under way to review and assess development approaches for the next generation of advanced reactors and to articulate a nuclear R&D strategy for the United States. In the broad context of alternative energy R&D strategies for reducing emissions of GHGs, however, an international study should be undertaken by DOE on criteria for globally acceptable reactors.

Such a study should be completed in no more than 5 years and should involve researchers and policymakers from around the world. Developing countries should be among the main participants. The study should attempt to establish criteria for an acceptable "global" nuclear reactor (or reactors) taking into account often conflicting requirements such as safety, reliability, scale, simplicity and standardization, waste disposal and storage, diversion-resistant fuel cycle, cost considerations, and fuel efficiency. Some of these requirements are briefly highlighted in Table 4-3.

Fusion

In the case of magnetic fusion, the United States ought to enter into international partnership arrangements for present and future R&D activities. Moreover, important questions could be better answered by a greater emphasis on fundamental physics rather than the current mixed emphasis on science and equipment development. A shift in DOE's focus toward much more basic research and a greater level of international collaboration will enable reductions in expenditures for the magnetic fusion program. The R&D funds so released ought to be reallocated among the other priority programs recommended in this study. In making this recommendation the committee does not preclude a resurgence in the scope and pace of funding for magnetic fusion RD&D should they be warranted by breakthroughs arising from the fundamental research program.

The committee recognizes that the foregoing recommendation is contrary to that made by an earlier NAS study,[15] that recommended "an increase over current funding of 20 percent, held steady for the next five years, followed by an additional increment of 25 percent to permit construction of the Compact Ignition Tokomak, resolution of the central scientific feasibility question and participation in the construction of an international test reactor." That recommendation, however, was offered as an interim one pending an assessment by the federal government of a "balanced array of technological alternatives as an insurance strategy for meeting U.S. and

TABLE 4-3 Criteria and Issues for the International Study on Advanced Reactors

Issues	Lines of Investigation
Safety and reliability	A demonstrable enhancement in safety is necessary. Minimum nuclear plant standards for siting and the protection of public health and environment should be defined in this global context. The optimal mix of passive versus active safety features should be explored.
Scale, simplicity, and standardization	The main attributes here are the scale of reactors, ease of construction, factory fabrication, and ease of operation and maintenance.
Nuclear waste disposal and storage	This issue is a critical public acceptance one and might conflict with requirements for diversion resistance. Minimizing the actinides in the waste stream is an interesting technical approach.
Diversion-resistance	Technical and institutional approaches should be explored for preventing illicit diversion of weapons related materials with world-wide deployment of nuclear reactors.
Cost considerations	Any comparison of relative costs with fossil-fueled systems should reflect the externalities usually ignored in traditional economic evaluations. Total and specific investment costs of advanced reactors should be examined. A 5-year construction time should be a worldwide target.
Fuel efficiency	This should be addressed in the worldwide context of different fuel cycles, resource needs and constraints, waste management, diversion-resistant criteria, and total system costs.

global long-term needs," including an assessment of the attributes of those alternatives (environmental, public health, safety, etc.). In the context of the current study, addressing alternative energy R&D strategies to deal with global climate change, the committee's view is that commercially viable fusion reactors are highly unlikely to make any significant additions to the U.S. electricity generation mix before the year 2050.

Review of DOE Renewables R&D Program

- Hydro

No ongoing R&D

- Biomass

Most relevant for electricity production are the DOE short rotating intensive culture program, started in the late 1970s and focused on fast-growing hardwoods, and the more recently initiated herbaceous energy crops program.

- Geothermal

The $18 million per year DOE program is aimed at developing drilling techniques and reservoir estimation.[11]

- Wind

The $9 million per year DOE wind program is aimed at developing proof-of-concept, variable-speed wind turbine systems and advanced airfoils and at better understanding of atmospheric fluid dynamics, aerodynamics, and structural dynamics.[11]

- Solar Thermal

The $15 million per year DOE solar thermal program is aimed at providing support for both the parabolic dish/dish-mounted engine generator and heliostat/central receiver concepts.[11]

- Photovoltaics

The $35 million per year DOE photovoltaics program includes thin-film polycrystalline and amorphous semiconductor research, high-efficiency crystalline materials research, fundamental and supporting research, and collector and system research.[11]

There is practically no ongoing DOE research on hybrid renewables/natural gas strategies. Ongoing DOE renewables-based hydrogen research is directed to investigations of very long-term possibilities involving photochemical, photobiological, and photoelectrochemical approaches.

During the 1980s enormous technical progress was made for a number of renewables during a time of minimum DOE R&D funding. A lesson from this experience is that government should resist the temptation to pick winners; this lesson was brought home by the commercial success with the parabolic trough thermal electric technology, despite the fact that the government's R&D effort has focused instead on the heliostat/central receiver and parabolic dish/distributed engine generator concepts.

For promising, rapidly evolving technologies such as photovoltaics, that are not yet ready to "take off" commercially, a prime consideration is how best to help sustain the industrial effort until significant commercial sales generate revenues sufficient to support the needed continuing R&D. A shortcoming of the federal program is its relatively weak analytical capacity for assessing potential future roles for renewables.

Renewables R&D Needs

- Hydro

The national potential for increasing hydro output at existing facilities should be analyzed.

The extent to which new technology can improve the economics of the untapped hydro potential and reduce the impacts on rivers needs to be defined.

An analysis is needed of the benefits to the United States of free-flow turbine technology for flowing river applications and equipment for unconventional small-scale hydro sources.

- Biomass

A program aimed at developing biomass-based power generation would emphasize biomass production, harvesting, and preprocessing; demonstrating biomass gasifier/gas turbine power technologies; and understanding better the long-term biomass resource base.

For near-term applications, R&D is needed on low-cost approaches for recovering biomass residues that are not now recovered. For longer-term applications, continued support is needed for the ongoing DOE program to grow terrestrial (woody and herbaceous) energy crops. R&D is also needed on biomass fuel processing problems—including innovative techniques for drying the fuel and densifying it (if necessary for gasification).

Considerable attention should be given to understanding better the potential for large-scale biomass-for-energy development in the United States, addressing not just the technical, economic, and environmental issues of biomass production but also the institutional and economic challenges of organizing the use of land for these purposes.

- Geothermal

Demonstration projects where continual engineering improvements can be evaluated, tested, and applied should be developed to obtain cost reductions.

- Wind

R&D is needed to increase energy recovery rates and to reduce capital costs. Collaborative programs between manufacturers and users are essential.

- Solar Thermal

In the case of the heliostat central receiver concept, further developmental efforts should build on the experience of the ongoing Phoebes project (i.e., the Jordanian/European/American joint venture to demonstrate a 30-MW system). In the case of the parabolic dish/engine generator concept, an expansion of the ongoing developmental effort should be made contingent on obtaining a major industrial champion.

- Photovoltaics

Resources should be committed to collaborative government/industry projects. The highest priority in this area should be given to generic module development and manufacturing processes.

A significant effort is needed on photovoltaics materials and cell fundamentals. Resources should continue to be committed to high-risk/high-potential payoff options involving new materials.

- Systems Analyses for Renewables

DOE should launch a new analytical effort, in partnership with both the photovoltaics and utility industries, aimed at assessing potential roles for renewables in the power sector over time. This systems analysis should be aimed at identifying market entry and strategic development paths and conditions under which large-scale penetration of the power sector by renewables is feasible—including various renewables/storage (including hydrogen) strategies, hybrid renewables/natural gas strategies, and renewables/strengthened utility grid strategies.

Current Transmission and Distribution (T&D) and Storage R&D Programs

Electric T&D and storage R&D programs are currently being carried out by DOE and the Electric Power Research Institute.

The DOE's FY 1990 appropriations for Electric Energy Systems and Energy Storage programs was about $30 million with $12.0 million directed for storage systems.[11] The budget request for FY 1991 of $40 million appears to reverse the trend of DOE's declining budget in these programs.[11]

DOE's R&D goals and budgets are directed toward system efficiency improvements and hence will help lessen the use of CO_2-emitting fuels. The goals should also improve system economics regardless of greenhouse effect considerations.

T&D and Storage R&D Needs

R&D programs pertaining to T&D and storage should initially focus on conceptual studies to examine and shape the most effective R&D program strategies. The analysis of T&D storage and dispatch issues will be interactive with the formulation and evolution of alternative generation approaches. It is important to adopt an integrated approach in the formulation of T&D and storage R&D programs. A comprehensive set of studies that consider the multiplicity of criteria involved in affecting R&D program formulation would represent a sound up-front investment before any expensive hardware projects are initiated.

The major strategies for T&D involve efficiency improvements, advanced communication and control technologies, advanced network system management, superconductor transmission lines, and energy storage technologies. The energy losses in T&D are approximately 8 percent of the energy supplied; therefore, 8 percent is the upper bound of direct energy savings that can be obtained by T&D system improvements. Indirect savings in CO_2 emissions by effective environmental dispatch of the generating system (to make maximum use of the non-CO_2-emitting generating equipment) may yield greater gains in overall CO_2 emission than the savings in direct energy losses in the T&D equipment.

Summary

There are many approaches for the electricity sector in the United States to achieve significant reductions in the emissions of CO_2 relative to the current baseline. The following are the committee's major findings for the various supply strategies considered that should shape an R&D program to address GHGs produced by electric power production. These are highlighted below under the generic categories of primary energy sources used in the electricity sector. No priorities are intended by the order of presentation; it simply reflects the order in which resources are addressed in this chapter of the report.

Although supply strategy options are primarily considered here, minimizing emissions of GHGs is also critically dependent on achieving high conversion efficiencies both in the electricity produced per unit of primary energy expended and in the services provided by each unit of electricity consumed at the point of use. Increasing end-use efficiency is a particularly important component of any global strategy to reduce and stabilize GHGs in the atmosphere.

Fossil Fuels

- Achieving substantial reductions in global GHG emissions will severely limit the use of coal as a primary energy source for electric power production unless economically acceptable means can be found for CO_2 removal and sequestering.

- For the near term, increasing the efficiency of fossil generating equipment is essential. The gas turbine/steam turbine combined cycle and fuel cells are currently available high-efficiency options. Substantial further improvements in the combined cycle, other advanced gas-turbine-based technologies, and fuel cells are possible. The systems driven by natural gas offer options for reducing GHG emissions.

- A priority of the coal R&D program should be to ascertain if there are economically and environmentally acceptable approaches for removing and sequestering CO_2. If such approaches can be found, coal conversion R&D priorities should be made consistent with their adoption.

- The Clean Coal Technology RD&D program has focused on reducing SO_x and NO_x emissions (i.e., acid rain). The application of clean coal technology in its current configuration could increase GHG emissions per kilowatt-hour produced. If uncorrected, the goals of acid rain reduction and GHG reduction may therefore be at odds. The Clean Coal Technology program should include and emphasize RD&D for high-efficiency conversion of clean coal to electricity.

Nuclear Energy

- Nuclear energy could be a major option to achieve significant reduction in CO_2 emissions in the electricity sector. It requires that the cost of the technology be reduced and its safety increased, if public acceptance of the option is to be achieved. Such acceptance is essential to permit the expansion of nuclear energy in the United States and on a global basis, as might be needed for an effective response to reducing CO_2 emissions.

- A nuclear reactor or reactors that can penetrate the global market for both developed and developing countries could be important for limiting CO_2 emissions in the mid and long term. The United States should take the lead in establishing an international study on the criteria for globally acceptable reactors (see Table 4-3).

- Magnetic fusion as a technology option for electricity production is many decades away from realization. The U.S. magnetic fusion program should focus on basic research with greater international collaboration.

Renewables

- Of the renewable technologies that could be available in the near term, biomass grown renewably and used to produce electricity in gasifier/gas turbine technologies offers potential as an option for a CO_2 balanced energy strategy.

- Significant advances have been made in improving the efficiencies and lowering the costs of photovoltaics, wind, and solar thermal technologies. The industry is currently serving niche markets but has not had the financial strength to support major efforts in manufacturing technology development. Small but steady purchases within the United States and federal assistance in international marketing would continue the significant technical advances and help market adoption of products that have been developed in the last 10 years.

T&D and Storage

- An improved and expanded T&D system could be an important option for making maximum use of non-CO_2-emitting technologies.

- New and improved technology for alternating current and direct current systems components is an essential part of developing an efficient, flexible, and reliable network needed to operate the electric power system in the most environmentally acceptable way.

- Energy storage capability is important to enhance the viability of intermittent renewables (e.g., wind and solar technologies) and enables greater efficiencies to be realized in the electricity delivery system in general.

TRANSPORTATION

Energy Use and GHG Emissions

In 1987 transportation consumed 22 quads (10^{15} Btu) and emitted about 440 MTC (as carbon dioxide) into the atmosphere. Four petroleum-based liquid fuels—gasoline, diesel, jet fuel (kerosene), and resid—account for about 95 percent of the U.S. transportation energy. The remaining 5 percent comes primarily from natural gas and electricity.

Table 4-4 shows how the 22 quads of transportation energy break down by mode of transportation and how much carbon was emitted by each combination of mode and fuel type. Cars and light trucks, fueled principally with gasoline, account for 57 percent of the transportation energy consumed in the transportation sector and also for an equal percentage of the carbon emissions; they are therefore the primary focus of this section. The three fuel-mode combinations that total 25 percent of the sector energy consumption and CO_2 production are also briefly addressed.

- Heavy gasoline trucks account for 5 percent of the transportation energy. They are not the subject of direct R&D recommendations but will benefit indirectly from much R&D directed at cars and light trucks.

- Diesel trucks, buses, and off-highway vehicles account for 12 percent of transportation energy. The diesel engine is an efficient power plant, but has NO_x and particulate emission problems that could limit its future use.

- Jet-fueled civilian aircraft account for 8 percent of transportation energy. Design of these aircraft is already heavily driven by the need for fuel efficiency, because of the impact of engine fuel consumption on aircraft range and direct operating cost.

For purposes of discussion, light trucks and passenger cars are generally combined; yet significant differences exist between the two. The sales of light trucks are about one-third the total light vehicle sales and are growing faster than passenger car sales. The average lifetime of light trucks is longer than that of passenger cars. Fuel economy standards for light trucks are less stringent than for passenger cars. These trucks, of course, often carry payloads but just as often are used for passenger transportation only. From the standpoint of fuel consumed and carbon emitted, light trucks are becoming increasingly important. Technologies in light trucks are very similar to those in passenger cars. Light trucks need special attention with regard to design of incentives that recognize their mission orientation and yet do not allow an underregulated vehicle category.

TABLE 4-4 Energy Consumption and Carbon Emissions by Mode in the U.S. Transportation Sector, 1987[16]

Combination of Mode and Fuel Type	Energy Consumed		Carbon Emitted[a]
	10^{15} Btu	% of Total	MTC
Primary Focus			
Gasoline cars and light trucks	12.56	57	254
Briefly Addressed			
Heavy gasoline trucks	1.17	5	24
Diesel trucks, buses, and off-highway vehicles	2.61	12	53
Jet-fueled civilian aircraft	1.85	8	37
	5.63	25	114
Not Considered			
Military aircraft	0.45	2	9
Water, gasoline + diesel + resid	1.33	6	27
Pipeline, natural gas + electricity	0.78	4	12
Rail, diesel + electricity	0.50	2	10
Other	0.78	4	16
	3.83	18	74
Total	22.02	100	442

[a]Estimated in this study. The emissions of carbon dioxide from the combustion of petroleum-based fuels are expressed in this table as million metric tons of carbon (MTC).

CFC Considerations

Air-conditioning equipment used in the transportation sector is a significant source of chlorofluorocarbons (CFCs), also potent GHGs. The Montreal Protocol has given greater impetus to use substitutes that are much less damaging environmentally than CFCs, to recover and reuse CFCs from discarded automobile air conditioners and other refrigeration equipment, and to conduct research and develop new working fluids that are environmentally benign.

Major Targets for Attention

Automobiles and Light Trucks

Because they produce 57 percent of the CO_2 emissions of the transportation sector, gasoline-fueled cars and light trucks are by far the most important target for emissions reduction. Two effective ways have been identified to reduce CO_2 emissions from cars and light trucks over the near term:

- improve vehicle fuel efficiency (or fuel economy), including increased use of smaller vehicles, and

- Use transportation systems more efficiently—by increasing automobile and light truck passenger load factors, switching from less efficient to more efficient transportation modes, and using existing modes more efficiently.

Improving Vehicle Fuel Efficiency. The technologies to improve new car and light truck fuel efficiency over the next 10 years are limited to those already in hand and those that are nearly ready for commercial application.[17] Radical innovation in mass-produced automobiles and light trucks is not possible in such a short time. In the United States new cars currently have an average fuel efficiency of about 28 miles per gallon (mpg). The fuel efficiency of the automobile fleet is about 19 mpg. Thus if no further improvements are made, by the year 2000 the U.S. automobile fleet will have a fuel efficiency approaching 28 mpg. Clearly, though, at least modest improvements in new cars are almost sure to occur; industry sources suggest that by the year 2000 new car fuel efficiencies will be about 32 mpg. With additional government policy actions, experts outside the industry believe that new car fuel efficiencies of 45 mpg are practical by the year 2000. These higher efficiencies would be achieved by greater improvements to cars and light trucks than those envisioned by the industry and by changes in the mix of cars purchased each year to more fuel efficient ones. Such policy actions might encompass new car fuel efficiency standards, higher gasoline taxes, taxes on fuel-inefficient cars ("gas guzzler" taxes), and rebates or tax credits on superefficient cars ("gas sipper" incentives).

The corporate average fuel economy (CAFE) standards enacted into law in the mid-1970s are credited by some experts with spurring the dramatic increases in fuel efficiency of new cars sold in the United States. Others, principally from the automotive industry, disagree, pointing instead to a period of higher gasoline prices and growing foreign competition. As of 1989, however, prices had dropped in real terms to preembargo levels, and much of the foreign competition revolved around vehicle quality and price.

Most experts believe, and economic analysis and European experience both suggest, that higher gasoline taxes are not likely to influence new car purchase decisions strongly toward more fuel efficient cars.

Gas guzzler taxes have been assessed since 1978, but their effects are obscured by the concurrent applications of fuel efficiency standards. There is no experience with gas sipper incentives. Many other incentives and disincentives can be envisioned, but their effectiveness is hard to predict.

Using Transportation Systems More Efficiently. Load factors in automobiles used for personal transportation are estimated to be 1.7 passengers per vehicle on average. Even modest improvements in this factor would substantially reduce vehicle miles traveled, with consequent reductions in fuel consumed and carbon dioxide emissions.

For example, as shown in Table 4-4, cars and light trucks consumed 12.6 quads of fuel in 1987 under the average load factor of 1.7 passengers per vehicle. If the load factor had been 2.5 instead, the fuel consumed would have dropped to about 8.6 quads; carbon dioxide emissions would have dropped from about 254 MTC to about 173 MTC. Increasing load factors, however, will not be stimulated by vehicle technology but by public policy. A wide range of policies could be considered to induce such changes. Among the more obvious are increased use of high-occupancy vehicle lanes, in congested areas, to speed travel of cars with several occupants; employer-supported (or -required) car and van pooling; substantial fuel taxes; and tailored parking fees and locations to encourage car pooling and discourage low load-factor use of cars.

Encouraging people to use public transit systems is another strategy for increasing the efficient use of transportation systems. Some of the public policies listed above, notably higher fuel taxes and higher parking fees (at destinations), would encourage greater use of mass transit. Other public policies include more and better park-and-ride facilities, free parking at mass transit facilities, and improved feeder systems to mass transit facilities.

However, only a small part of the demand for personal transportation in the United States is satisfied by mass transit systems; so even if these were used far more extensively than they are now, little energy would be saved and carbon emissions would not be greatly reduced. For example, illustrative calculations suggest that if the use of mass transit could somehow be tripled and travel in personal vehicles reduced correspondingly, only about 10 percent of the energy used for land transportation of people would be saved. Nevertheless, other good and sufficient reasons may exist for supporting mass transit (e.g., to reduce local air pollution and congestion and encourage reductions in urban sprawl).

Near Term. It is reasonable to expect some reduction in GHG emissions from the personal transportation subsector by the year 2000. Table 4-5 shows what might be accomplished.

The base case for 1987 is taken from Table 4-4. Scenario 1 allows for improvements in fleet fuel efficiency from 19 mpg to 28 mpg and a 20 percent growth in passenger miles traveled (without a change in load factor). Fuel consumption would drop from 12.6 to 10.3 quads, with a proportional reduction in carbon emissions from 254 to 208 MTC.

Scenario 2 takes into account a 30 percent growth in load factors. This would provide further reductions of 2.3 quads in energy consumed and 47 MTC emitted.

Scenarios 3 and 4 take into account the effects of more ambitious fuel efficiency standards that might raise fleet averages to about 32 mpg by the year 2000.

While these calculations are intended only to be illustrative, they suggest that public policy can reduce emissions of GHGs if there is the determination to do so.

Long Term. Should it be necessary in the post-2000 period to make massive reductions in carbon emissions from the transportation sector, far more drastic changes than those outlined above will be required. In addition to the development and manufacture of more efficient vehicles and the more efficient use of vehicle fleets and systems, new energy sources (or fuels) must be found for vehicles. For cars and light trucks the three most promising possibilities are

- alcohol fuels (ethanol and methanol) made from biomass in a fuel production system that absorbs in biomass production at least as much carbon as is emitted in producing and consuming the alcohol fuels;

- electrification of cars and light trucks, with the electricity coming from nuclear or renewable-fired generating plants; and

TABLE 4-5 Energy Consumption and Carbon Emissions from Automobiles and Light Trucks: Illustrative Scenarios for the Year 2000

Scenario	Load Factor	Fleet Average Fuel Economy (mpg)	Passenger Miles Traveled*	Energy Consumed (10^{15} Btu)	Carbon Emitted (MTC)
Base 1987	1.7	19	–	12.6	254
Scenario 1	1.7	28	1.20	10.3	208
Scenario 2	2.2	28	1.20	8.0	161
Scenario 3	1.7	32	1.20	9.0	181
Scenario 4	2.2	32	1.20	7.0	140

* relative to 1987

- hydrogen produced, for example, by electrolysis of water with electricity generated without emissions of GHGs.

Cars and light trucks, fueled with electricity or hydrogen, are likely to be limited in range and/or performance compared to conventional vehicles. Vehicles using alcohol fuels will be more nearly comparable to current models but will require larger fuel tanks to achieve equal range. Moreover, large-scale use of any of these alternative fuels implies a massive developmental and industrial effort to create new fuel production plants; electricity generating plants; and, in the case of electric cars, production capabilities for advanced storage batteries.

The real possibility must also be taken into account that alternate fuels will not develop into viable means for meeting the massive needs for personal mobility in the first half of the next century. There may be higher-value uses for such alternate fuels. Thus, fossil fuels would remain the only option. Such possibilities reinforce the importance of vigorous long-term pursuits of the near-term objectives stated above: technologies to enable development of very efficient vehicles and means to increase the efficient use of transportation systems.

Other Modes of Transportation

In this section other modes of transportation are briefly discussed, including diesel-powered vehicles, principally larger intercity freight trucks; large gasoline-powered trucks, used principally to haul freight within cities; and airplanes, principally commercial jet passenger aircraft.

Diesel Trucks. Diesel trucks, largely "18-wheelers" used to haul freight over the interstate highway system, annually consume about 2.4 quads. Fuel costs are currently a significant part of the operating costs of these vehicles, and strong economic incentives exist to make these trucks more efficient and productive. These incentives are evidenced by larger more aerodynamic trucks, many with improved engines, that have been appearing for some time on the nation's roads.

While the committee did not consider mode switching in much depth, freight trains are far more efficient than trucks. It has been estimated that freight can be transported by rail for about 500 Btu per ton mile, versus 3,400 for heavy trucks. Such mode switching could be encouraged by proper economic incentives and perhaps by development of improved systems to move truck trailers by rail—for example, road-railers, trailers that permit easy conversion from road to rails and back.

Fuel switching might be more promising for diesel trucks than for cars. Alcohol fuels (produced from biomass in a carbon balanced system) would be the fuel of choice; diesel engines, for example, can use methanol with relatively minor changes. Moreover, fuel tank capacity in diesel trucks can be increased more easily than in cars and light trucks to offset the lower energy density of the alcohol fuels. Thus, if alcohol fuels can be produced in a carbon-balanced cycle, heavy trucks are promising applications for such fuels.

Large Gasoline-Powered Trucks. Gasoline-powered trucks serve mainly, though not exclusively, to move freight within cities. There are few alternatives to trucks to move such freight, and not much improvement in fuel efficiency is likely to be achieved given the nature of the service these vehicles provide. Some electrification and conversion to diesels can, of course, be carried out. But, as in diesel trucks, the most promising avenue is to switch from petroleum- to biomass-based fuels.

Commercial Passenger Aircraft. Commercial aircraft consume about 1.85 quads of fuel; virtually all of this is used by commercial passenger airplanes. General aviation and air freight account for about 0.2 quads at most.

Modern subsonic jet airframes have been carefully optimized to reduce fuel consumption. Both technical and economic considerations impel such optimization, since both aircraft range and cost per seat mile are important measures of aircraft performance. Aircraft power plants have received similar attention, and at regular intervals jet engine manufacturers introduce engines with improved fuel efficiency. Continuing improvements are being made in lighter structural materials for aircraft and higher-temperature materials for jet engines; thus, improvements in the energy efficiencies of commercial airplanes are likely to continue. Both the development of airframes and engines are supported indirectly by the U.S. Department of Defense's efforts on improved materials, engines, and aerodynamics for military airplanes and by the National Aeronautics and Space Administration through its continuing program of aeronautical research.

The energy intensity of air travel, as measured in British thermal units per passenger mile, declined at a rate of about 4.[illegible] percent per year from 1972 to 1985.[18] Such trends toward more fuel-efficient airplanes are likely to continue, but at a somewhat slower pace because the technologies are mature and equipment turnover rates are relatively low.

In the near term, fuel switching for airplanes is not feasible. Even in the long term—out to the year 2050—fuel switching on any substantial scale is unlikely. The most likely candidate fuels are liquefied methane and liquefied hydrogen, but both are cryogenic and present serious safety, technical, and economic challenges.

Achieving Major GHG Emissions Reductions in the Long Term

Table 4-6 examines how the transportation sector might look by the middle of the next century. The table is based on many debatable assumptions and is intended only to illustrate the problems associated with major reductions in GHG emissions from this sector.

The column labeled 1987 shows estimated carbon emissions for that year from Table 4-4. The next column illustrates what might happen without any fuel switching. It is assumed that by the year 2050 the U.S. population will have reached a level 50 percent above that of 1987, with no change from the 1987 average miles traveled per person. This means that passenger miles traveled in 2050 will be 50 percent more than in 1987, equivalent to an average annual growth rate of about 0.65 percent. The fleet average of personal vehicles is assumed to be 50 mpg of gasoline. Car and light truck load factors were placed at 1.7 to 2.5 to calculate the range shown. Other major demands for transportation, driven principally by economic growth, were assumed to increase their energy consumption by 50 percent between 1987 and 2050, as the result of increased demand for service, partly offset by increased efficiency. Such a scenario would lead to carbon emissions of 360 to 410 MTC in the year 2050.

By substituting biomass-derived fuels—principally methanol—for diesel fuel in the applications likely to be most amenable to such substitution, carbon emissions might be reduced by about half, to the 190—240-MTC range. Some 9 quads of methanol, however, or about 9,000,000 barrels per day, would be required to effect this substitution.

By further improving personal vehicle fuel efficiency to 100 mpg, emissions could be reduced to 140 to 170 MTC. Finally, by substituting biomass-derived fuels, hydrogen, or electricity for petroleum-based fuels in personal transportation vehicles, carbon emissions might be reduced to about 100 MTC, or about 20 percent of what they might be without aggressive efficiency improvements and fuel switching.

Further significant reductions could be accomplished only by applying new fuels such as methane and hydrogen to aircraft.

TABLE 4-6 Carbon Emissions from the Transportation Sector: Four Illustrative Scenarios for the Year 2050[a]

	Carbon Emissions MTC				
Mode	1987	Scenario 1 2050 (50 mpg)	Scenario 2 2050 (50 mpg)	Scenario 3 2050 (100 mpg)	Scenario 4 2050 (100 mpg)
Cars and light trucks	254	98[b]-145[c]	98[b]-145[c]	49[b]-72[c]	0
Diesel trucks	53	80	0	0	0
Gasoline trucks	24	36	0	0	10
Aircraft	37	56	56	56	56
Water transport	27	41	0	0	0
Pipelines	12	12	12	12	12
Railroads	10	15	0	0	0
Other	16	16	16	16	16
Military	9	9	9	9	9
Total	442	363[b]-410[c]	191[b]-238[c]	142[b]-165[c]	103

[a] Illustrative calculations made in this study. The four scenarios are as follows: Scenario 1, Base case for the year 2050-no biomass-derived fuels, modest gains in auto and light truck fuel efficiency; Scenario 2, Biomass fuels and electricity introduced into the most amenable services; Scenario 3, Same as scenario 2 but with greatly improved fuel efficiency of personal passenger vehicles; Scenario 4, Same as scenario 3, but with biomass, electricity, or hydrogen introduced into personal passenger vehicles.

[b] Load factor = 2.5.

[c] Load factor = 1.7.

R&D Priorities in Transportation

Vehicles and systems for the transportation sector are produced by large, technically competent, well-capitalized firms in the United States and abroad. Most if not all of these firms carry out extensive R&D on new products and maintain relatively high rates of market-driven innovation in their cars, trucks, aircraft, aircraft engines, and diesel engines. Fuel efficiency, though, has not been a high priority in the research programs of the automobile companies except when they have been spurred by fuel efficiency standards. Thus, without such standards or other incentives, progress on fuel efficiency should not be expected from these companies. Nevertheless, federal R&D can add little of value by allocating resources to technologies nearing readiness for commercial application. Thus, the federal R&D role should encompass the three broad classes of activity to which private interests are unlikely to devote much effort:

- research aimed at strengthening the technology base in technologies of special relevance to transportation (e.g., high-temperature materials for engines and light structural materials for vehicle structures);

- development of innovative vehicle components and concepts to understand the major technological and engineering issues they present and to permit preliminary assessments of their potential value; and

- studies to assess the likely effectiveness of policies to effect changes in the transportation systems of U.S. cities and regions that would lead to reductions in GHG emissions in this sector.

Technology Base Research

The following areas are especially deserving of attention:

- combustion, especially that relating to the most efficient utilization of alternative fuels in spark ignition and diesel engines, and including work to reduce regulated emissions to meet projected standards;

- high-temperature structural materials such as advanced ceramics and metals to permit more efficient engines;

- economical lightweight composite structural materials to reduce vehicle weight;

- structural materials applicable to engines, transmissions, and load-bearing parts;

- computational aerodynamics and fluid mechanics addressing vehicle drag reduction and improvement in components such as torque converters; and

- tribology, aimed at reduction of engine and drive train frictional losses.

Vehicle Components and Systems

The following vehicle components and systems are recommended for research:

- strong effort focused on battery concepts that show the most promise;

- hybrid vehicles, including combinations of small heat engines and energy storage devices;

- onboard storage mechanisms for alternative fuels, hydrogen and methane, and distribution and storage systems for hydrogen;

- onboard fuel cells for generation of propulsive electric power;

- onboard photovoltaics for accessory power;

- engine systems to achieve better integration of engine, transmission, and ancillary components to improve part-load fuel economy;

- optimization of engines and vehicles, for alternative fuels, including work on materials compatibility with methanol and ethanol;

- vehicle system studies to address such issues as construction of crash-worthy cars of light structural material, control of emissions in superefficient engines, hybrid configurations that permit continuous full-load operation of internal combustion engines; and

- new innovative systems approaches to major transportation problems (e.g., electrified highways to couple, inductively or otherwise, electric vehicles to external sources of energy).

Policy Studies

Continuing efforts should be undertaken by the federal government to

- assess the technological potential to improve automobile fuel efficiency while preserving other important attributes of cars such as safety, comfort, performance, and costs;

- identify and analyze new incentives to effect fuel economy improvements, load factor growth, and mode shifts;

- analyze ways to phase in a substantial gasoline tax while using revenues to offset its regressive impact;

- monitor results of policies (e.g., South Coast Air Quality Management District) adopted to encourage car pooling and the use of alternative fuels and electric vehicles; these results could well have application in the future to policies that might be adopted nationally; and

- monitor the effectiveness of actions to control CFCs.

An Assessment of DOE Transportation Research

The transportation R&D program, under DOE's Office of Conservation, includes the following programs that are pertinent to GHG emissions:[11]

Program	FY 1990 Funding (millions of current $)
Automotive gas turbine	$12
Low heat rejection diesel	5
Electric vehicle battery R&D	8
Electric vehicle propulsion	6
Advanced materials	15
Total	$46

In the committee's estimation, the automotive gas turbine engine is unlikely to become the power plant of choice for cars or trucks, unless (in the case of heavy trucks) diesel engines are not able to comply with the kind of regulated emission standards that are likely to be developed in the future.

Thus, the vehicular gas turbine engine should be considered a backup power plant option. At this time it does not merit further support for engine development. Nevertheless, engines that have already been developed under the Advanced Turbine Technology Applications program should continue to be used as test beds for evaluating ceramic materials. This role is merited.

The low heat rejection diesel engine deserves more attention than it is currently getting. Its application in both heavy and light trucks should be explored. However, waste heat recapture should not be investigated until the question of whether ceramics can be run successfully in the diesel engine is resolved.

With regard to electric vehicles, the committee believes that the main barrier appears to be the limited capability of batteries to store and deliver energy, and that is where the research should be performed. DOE should assess the number of different battery types being worked on and eliminate those that do not survive a strong unbiased assessment. Propulsion work should be deferred until the battery problem is solved.

Advanced materials research, including ceramics, can enhance energy efficiency of a number of applications and is endorsed by the committee.

RESIDENTIAL AND COMMERCIAL BUILDINGS

Energy Use and GHG Emissions

The buildings sector in the United States is highly diverse, consisting of single-family houses and a variety of multifamily housing units in the residential subsector and office, retail, restaurant, hospital, hotel, warehouse, school, and other construction types in the commercial subsector. Within each of these building types is a wide range of sizes, energy loads for heating and cooling, ventilation, lighting, installed equipment, and occupancy. Finally, these buildings are located in areas of vastly different climates.

In addition to this diversity in use and climate, technologies are developed, deployed, and operated in this sector that involve a wide variety of organizations—custom home builders to tract developers, architects, design engineers, appliance and mechanical equipment manufacturers, construction firms, lighting engineers and manufacturers, and building owners and managers. The number of decision makers involved in energy-related issues in the buildings sector approaches the total population of building users from homeowners and tenants to shopkeepers, office workers, and building managers. These factors add considerable complexity to the development and implementation of energy-efficient technologies in this sector.

Table 4-7 summarizes current annual energy use in the buildings sector by major category of service demand and by fuel type for residential and commercial applications. Table 4-8 presents estimates of the total current emissions of CO_2 arising from this energy use, including electricity.

TABLE 4-7 Current Energy Use in the Buildings Sector [a]

Application	Energy Use 10^{12} Btu					
	Gas	Oil	Electric	Other	Total	Total Primary
Residential[b]						
Space heating	3450	950	750	800	5950	7450
Water heating	780	70	390	20	1260	2040
Air conditioning	10		390		400	1180
Refrigeration	0		570		570	1710
Cooking	260		170		430	770
Clothes drying	60		170		230	570
Clothes washing	0		20		20	60
Dishwashing	0		30		30	90
Other	0	500	450	70	1020	1920
Total residential	4560	1520	2940	890	9910	15790
Commercial[c]						
Space heating	1580	1000	590		3170	4350
Water heating	80	50	40		170	250
Space cooling	30		780		810	2370
Lighting	0		860		860	2580
Cogeneration	120		0		120	0[c]
Cooking	250		?		250	250[d]
Other	330		320	20	670	1310
Total commercial	2390	1050	2590	20	6,050	11,110
Total residential and commercial	6950	2570	5530	910	15,960	26,900

[a]Estimates from various sources for energy use in 1985-1988.[19,20]

[b]Total primary is the sum of oil, gas, and other energy use together with three times electric energy use.

[c]Total primary is calculated for residential above except for gas use in cogeneration, which is assumed to displace an equal amount of primary energy.

[d]Gas only

TABLE 4-8 Current CO_2 Emissions by the Buildings Sector[a,b]

Energy Use	CO_2 Emissions, MTC/year From Fossil Fuel Use	Attributable to Electricity Use
Residential		
Space heating	69.3	36.4
Water heating	12.7	18.9
Air conditioning	0.1	18.9
Refrigeration	—	27.7
Cooking	3.8	8.3
Clothes drying	0.9	8.3
Clothes washing	—	1.0
Dishwashing	—	1.5
Other	10.2	21.9
Total residential	97.0	142.8
Commercial		
Space heating	43.2	28.7
Water heating	2.2	1.9
Space cooling	0.4	37.9
Lighting	—	41.8
Cogeneration	1.7	—
Cooking[c]	3.6	?
Other	4.8	15.5
Total commercial	55.9	125.8
Total residential and commercial	152.9	268.6

[a]Calculation based on Table 4-7. CO_2 emissions expressed as millions of metric tons of carbon (MTC) per year. The conversion factors (in MTC/10^{15} Btu of fuel use) are as follows: natural gas, 14.5; oil, 20.3; coal, 25.1.

[b]Electric generation mix: 10% of natural gas, 5% of oil, 55% coal, and 30% non-CO_2.

[c]Gas only.

Major Targets for Attention

Table 4-9 presents the technologies and practices that in the committee's view will be the most important contributors to achieving large reductions in GHG emissions in the buildings sector, in both the near term and the long term.

TABLE 4-9 Potential Contribution of Building Technologies/Practices to Reduction of GHG Emissions[a]

	Residential		Commercial	
Technology/Practice	Near Term	Long Term	Near Term	Long Term
Energy Conversion Technologies				
High-efficiency gas heating	3	2	3	2
Advanced heat pumps	1	2	2	2
High-efficiency cooling	1	2	2	3
High-efficiency hot water	3	3	2	3
Solar hot water	1	2	-	-
Solar photovoltaics	-	2	-	2
Cogeneration	-	1	3	2
Energy storage	-	1	1	2
Commercial refrigeration	-	-	2	2
Building Components and Systems				
Controls	1	2	3	3
Advanced windows	2	3	1	3
High-efficiency lighting	1	3	3	3
Efficient residential appliances	2	3	-	-
Efficient office equipment	-	-	1	3
Insulation	2	2	2	2
Construction materials	-	2	-	2
Design/Practice				
Practice: data, construction, O&M[b]	3	3	3	3
Community design	1	3	-	3
Design for manufacturing, assembly and operation	1	3	1	3

[a] Rating scale is as follows: 1 - can make some contribution to GHG reduction, 2 - can make considerable contribution to GHG reduction, 3 - can make substantial contribution to GHG reduction.

[b] O&M, operation and maintenance

Path to Reduce Emissions

Reduced Energy Service Demand

A primary strategy for reducing GHG emissions is to apply new technology to reduce the demand for energy in buildings. The most effective means in the existing residential market sector is through building envelope retrofits. Both analysis and demonstration of envelope retrofits have shown savings on the order of one-third to greater than one-half.[21-23] Typical retrofit measures include increased wall, floor, attic, and duct insulation; window replacement or the addition of advanced-technology storm windows; and measures to reduce infiltration of outside air, including weatherstripping and the installation of outside door closers.

These results would suggest that the space heating load of older houses can be reduced by approximately 50 percent. For example, if 30 percent of the unimproved pre-1970 houses could be retrofitted, an energy savings of approximately 0.8 quads could be expected.

Envelope improvements that reduce space heat demand also contribute to reductions in space cooling requirements. However, the analysis in the above-referenced reports shows smaller percentage improvements in cooling demand, on the order of 10 to 30 percent. Nevertheless, these savings are electric peak power reductions, so they translate into significant primary energy savings and peak power savings, adding to their cost-effectiveness.

New technology for building envelopes will have little impact on the energy service demands of existing commercial buildings. However, the committee estimates that energy use can be reduced by at least 10 percent through improved operating procedures in buildings, such as optimal use and control of HVAC equipment, lighting, and other office equipment. Research on appropriate lighting levels for peak productivity and on the necessary ventilation level for indoor air quality may allow some small reductions in service demand in the near term. Energy-saving improvements can also be incorporated into existing commercial buildings when remodeling is done for new tenants.

Improved operating procedures in existing buildings can be brought about through the use of building automation and control technologies, in conjunction with better training of operating personnel. The building automation industry is now about 20 years old. Its development coincided with and was made possible by the evolution of computer technology. Today's microcontrollers make comprehensive control systems possible and practical for many existing buildings as well as for all new buildings.[24,25] Despite this technical capability, building control systems "are among the most underused and misunderstood devices used in operating a modern building."[24] Further R&D on usability of controls, as well as training and motivation, is necessary and will yield both future energy savings and GHG emission reductions.

Human control systems (shutting off lights, fans, etc., when not needed) could also have important impacts. Diagnostic and other feedback mechanisms can be developed to enhance such systems as well as to motivate appropriate implementation.

New building envelope technology can have a major impact on reducing the space-conditioning demands of future commercial buildings. Wall and building cladding technologies can reduce heat loss through walls by a factor of more than 50 percent. ASHRAE 90 (published by the American Society of Heating, Refrigerating, and Air Conditioning Engineers) establishes a standard for wall heat loss (U value) of 0.155 Btu/h/ft^2/°F for a climate of 5,000 heating degree days. Research is under way on evacuated silica aerogel insulation materials that achieve U values of less than 0.02 for a 3-in.-thick panel.[26]

Similarly, windows have already been designed that control both convective heat transfer and infrared emission. Windows with U values of 0.2 are already available;[27] R&D is under way that will reduce the Uvalue to 0.1 or lower if winter solar gain is considered. Roofing and ceiling technologies are on similar paths in terms of convective heat loss. Reflective roof materials and rooftop evaporative cooling systems are also available.

The possibility of constructing a building that requires virtually no energy to make up for convective or radiative gains/losses is a reasonable goal. Space-conditioning demands will be dominated by requirements for ventilating air and for removing heat generated by building equipment and occupants.

Increased Efficiency of Energy Conversion

All of the service demands of the building sector are supplied by energy conversion devices, such as furnaces, air conditioners, or light bulbs. In the area of residential space heating, technology is already available to significantly improve average heating efficiencies, and R&D is under way that can lead to even greater improvements. Table 4-10 shows the improvements achievable.

TABLE 4-10 Space Heating Equipment Seasonal Performance Factor (SPF)[a]

Technology	Current Mix	NAECA [b] Standard	Today's Best	Best Prototype
Gas furnace	0.6[c]	0.78	0.95	
Zoned gas furnace[d]	NA	NA	NA	1.2
Gas sorption heat pump	NA	NA	NA	1.7
Gas engine heat pump	NA[e]	NA	1.3[e]	1.7[f]
Electric furnace[g]	0.98	NA	0.98	–
Electric heat pump[g]	1.80	2.0	2.60[h]	–

NA, not available.

[a]Seasonal performance factor (SPF) is a measure of useful heat output divided by energy input averaged over seasonal temperatures and demand variations. SPF is equivalent to efficiency in simple combustion furnaces but can be much higher in heat pumps that extract energy from the outdoor air.

[b]National Appliance Energy Conservation Act of 1987.

[c]Based on a comparison of reported gas consumption for space heating in the American Gas Association's Househeating Survey and DOE's Residential Energy Consumption Survey versus modeled heating demand.[22]

[d]Modulating zoned gas furnaces achieve improved SPF by individual room control of heating requirements. This is not really an efficiency improvement but is a measure of reduced demand for space heat in a residence.

[e]Gas engine heat pumps are available commercially in Japan.[28] None are currently marketed in the United States.

[f]Prototype system currently in field test.[29]

[g]If electricity efficiencies were given in terms of source energy to generate electricity, it would be necessary to multiply these SPFs by 0.3 to 0.4.

[h]Two high-efficiency electric heat pumps have recently entered the market.[30]

The National Appliance Energy Conservation Act of 1987 (NAECA) standards take effect on January 1, 1992. Given an average life of approximately 20 years for a gas furnace, about half of the furnaces in place in the year 2000 should meet the 78 percent minimum efficiency standard. For the most part these will replace the oldest, lowest-efficiency furnaces in the current mix. Therefore, with no further action the national average gas furnace efficiency should reach about 70 percent by the year 2000.

A vigorous equipment replacement program could achieve a goal of installing 78 percent efficient furnaces in 60 percent of existing houses and 90+ percent efficient furnaces in 40 percent of existing houses. The result would push the national average gas furnace efficiency to 85 percent with a 35 percent energy savings compared to the current mix. Successful introduction of gas heat pumps would have even greater effects.

Similar impacts can be expected in electric heat pump energy use. If the average heat pump life (including the compressor) is close to 12 years, by the year 2000 almost all electric heat pumps will meet the minimum NAECA standards for heating.

Very significant reductions can be achieved through retrofitting existing heating and cooling equipment with existing technologies. R&D can be expected to result in continued increases in efficiency. Table 4-11 shows the efficiencies for installed capacity, currently available technologies and those on the drawing board.

A study conducted by the city of Phoenix showed that replacing existing equipment with the current state of the art as well as optimizing capacity (most installed equipment was significantly oversized) would result in average energy use reductions of 45 percent.[31] The study concluded that these retrofits would have payback times of 1 to 7 years; 12 of the 16 cases analyzed had payback times of less than 3 years.

Given the expected lifetime of 20 years for most HVAC systems, replacing at least 50 percent of existing systems with high-efficiency systems in the near term is reasonable. Success in such a retrofit program would reduce, in the near term, heating energy use in buildings by 10 to 15% and cooling energy use by 20 percent. Additional savings could be expected in the long term as a result of further retrofit of existing technology and successful R&D on the concepts shown in Table 4-11.

R&D on lighting technologies has already resulted in major improvements that are available but not in general use. These technologies include compact fluorescent bulbs installed in conventional incandescent sockets, improved incandescent and fluorescent lamps, fluorescent lamp ballasts, and improved fixtures. These technologies increase the lighting efficacy from today's standard light bulb level of 5-10 lumens/watt (lm/W) to 20-25+ lm/W for the best incandescent bulbs and 90-100 lm/W for the best fluorescent systems.[32,33,34].

Installation of high-efficiency lighting is also very cost-effective in most building applications.[32] R&D is already under way that has the potential to further increase lighting efficiency to 100 lm/W in the near term and ultimately to 200 lm/W.[35]

In the near term a 40 to 50 percent reduction in the energy used for lighting existing commercial buildings should be feasible. Improved lighting technologies provide a secondary benefit by reducing the building cooling load, offset slightly by an increased winter heating load.

Cogeneration technologies offer significant potential for energy and GHG emission reductions. As much as 40,000 MW of demand in the existing stock of buildings is estimated to be suitable for cogeneration.[36] Since cogeneration systems provide both electric and thermal power to a building, they displace a roughly equal amount of energy otherwise required for the thermal loads of the building. The thermal energy generated can be used year-round in many applications, especially if the system is integrated with a thermally driven absorption cooling system.

If gas-fired cogeneration were successfully installed in all of these potential applications, GHG emissions for the commercial building sector would be cut by 6 percent.

TABLE 4-11 Space Cooling Equipment SPF[a]

Technology	Current Mix	Today's Best	Best Prototype
Electric air conditioner	2.2	3.0	3.6
Electric heat pump	2.4	4.5	—
Commercial electric chiller	3.0	5.0	5.5
Gas air conditioner	0.5	0.6	1.0
Gas heat pump	NA	1.1	1.3
Gas absorption chiller	0.9	1.0	1.2
Gas engine chiller	NA	1.5	1.7
Gas engine chiller + absorption	NA	1.9	2.0

NA, not available.
[a] SPF defined as in Table 4-10.

Further improvements in efficiency and operation of refrigeration equipment is also possible. Direct and indirect evaporative (adiabatic) cooling and other retrofits can cheaply boost outputs by desuperheating hot gas entering condensers, precooling air entering air-cooled condensers, and subcooling liquid refrigerants leaving condensers.[37] Substantial improvements in the coefficient of performance (COP) of gas-fired sorption refrigeration systems are also possible. R&D is under way with the goal of increasing COPs from 0.6 to 1.0.[38,39]

Fuel Substitution

Gas technologies, even with lower SPFs, generate fewer GHGs than electric technologies because of the power plant emissions associated with electricity generation.[40] Replacement of electric resistance space and water heating represents a viable strategy for near-term reductions in GHG emissions. For example, the current residential water heating market is composed of approximately 38 percent electric resistance water heaters.[41] Replacement of one-half of these with 90 percent efficient gas water heaters would reduce GHG emissions from residential water heating by 25 percent. GHG emissions are also reduced by substitution of gas for oil. Nearly one-third of the commercial space-heating energy is provided by oil-fired systems. Replacement of half of these with efficient gas space-heating systems would reduce GHG emissions from commercial space heating by nearly 25 percent.

Solar water heating technology is a proven and effective means to reduce nonrenewable energy use by substituting renewable energy. If solar-assisted water heating, providing a 50 percent solar input, penetrated 10 percent of the existing residential market by the year 2000, water heating energy use would be reduced by 10 percent.

Impressive cost reductions have been made in photovoltaics that should soon make the systems viable for housing. Reductions in the thermal envelope, lighting, mechanical, and appliance electrical loads will help interface reasonably sized photovoltaic systems. Utilities should be encouraged to view roof areas as "prime real estate" for photovoltaic panels, which could offset daytime loads from industrial and commercial customers.

Gas-fired cooling systems are now available, and improved systems are under development. Engine chillers, absorption chillers, and desiccant-based systems are obtainable. Currently, such gas systems provide only a few percent of the cooling requirements of the commercial building sector. They offer several advantages, including ease of integration with cogeneration systems and (for desiccant or sorption systems) no dependence on CFC or other GHG-based refrigerants. If 20 percent of the commercial cooling needs were provided by gas-fired systems, GHG emissions for cooling existing buildings would be reduced by about 25 percent.

Summary

Table 4-12 summarizes the potential impacts of technology on reducing energy use and GHG emissions in the buildings sector.

TABLE 4-12 Potential CO_2 Reductions in the Buildings Sector[a]

	Percent Reduction[b] in CO_2	
	Near Term	Long Term
Reduce energy service demand		
Retrofit existing homes	2.9	0.9
Advanced construction of new buildings	5.6	24.7
Improved building operating practice	4.5	4.5
Increase equipment efficiency		
High-efficiency heating and heat pumps	8.7	17.8
Cogeneration	1.4	6.9
High-efficiency lighting	5.9	5.9
Use alternate fuels		
Replace half of oil heat with gas	3.1	3.1
10% penetration of solar photovoltaics	NA[c]	6.9
Gas/solar water heating	1.7	2.5
Impact of examples[d]	30.0	60.0
Other impacts		
25% Savings in other areas	7.0	NA[c]
50% Savings in other areas	NA[c]	14.0
Total potential impact	37.0	74.0

[a]Estimates based on analysis in this study by the Buildings Panel.

[b]Percent reduction from total CO_2 emissions (including those due to primary electric generation) for the residential and commercial sector per unit of service provided (i.e., per household or commercial square foot). Thus, these do not represent reductions from current levels since growth in the number of households and commercial square footage is not included.

[c]Either not applicable or not expected to have impact in time frame shown.

[d]Impacts are not additive. Obvious double counting (e.g., demand reduction and equipment efficiency) has been accounted for; other effects (e.g., cogeneration, solar, photovoltaics, electric) could result in some double counting, especially in the long term.

Over the long term, energy use and CO_2 emissions (per household for residential space and per square foot for commercial space) can be reduced by more than 70 percent. The long-term results must be viewed with the recognition that they are based on projected energy use and GHG emissions in the year 2000. Energy use patterns will likely be quite different then, especially if the near-term R&D and implementation actions recommended are successful. For example, since building space-conditioning energy requirements will be greatly reduced, energy use for office equipment, appliances, etc., will become more significant. This explains the growing importance of these technology areas in the long term, as was shown in Table 4-9.

R&D Needs and Priorities

Near Term

The most important near-term impacts will result from technologies that are already developed but that, for a variety of reasons, are not yet fully implemented. The highest-priority near-term R&D will focus on verifying the reliability and durability, cost-effectiveness, and performance of these technologies.

The following areas of R&D are considered most important to meet the near-term target for GHG emission reductions.

Energy Conversion Technologies. As Table 4-9 indicated, high-efficiency gas heating, high-efficiency water heating, and cogeneration are expected to provide the greatest impact on reducing GHGs in the near term.

While the greatest impacts will result from the application of technologies already developed and commercially available, this does not eliminate the need for R&D that further improves efficiency or is specifically targeted to the reduction of GHG emissions. Important examples in the energy conversion area include advanced heat pumps, gas cooling/refrigeration, alternate refrigerants, engine durability and performance, and thermal storage.

Building Components and Systems. Table 4-9 showed that the greatest near-term benefits are expected from advanced windows, appliances, and insulation in the residential sector; in the commercial sector the benefits are from controls and high-efficiency lighting. Although advanced technologies are available in most of these areas, R&D has the potential to produce improvements on the order of a factor of 2 in the case of lighting and windows.

In the case of residential appliances, very little has been done to develop products with reduced energy usage. Examples of important R&D needs include controls and electronics; low-energy appliances-cooking, clothes drying, dishwashing; non-CFC blowing agents for foam insulation; advanced windows—low-emissivity, gas-filled, improved frames; and advanced lighting.

Design/Practice. Perhaps the greatest near-term reductions in GHG emissions can be achieved through improved operating practices. Although this area must be addressed primarily through various implementation approaches, several areas of R&D can have significant impact. Furthermore, this is a neglected R&D area probably because it tends toward the social and behavioral sciences rather than the physical sciences or engineering. Examples of important R&D include the following:

- Decision-making methods—"expert" systems[42] for (1) selecting the most cost-effective HVAC and envelope technologies, (2) commissioning these buildings and systems to ascertain if the design is successfully installed, and (3) operating the buildings and systems as designed.

- Motivation and other social science research—identification of motivational approaches to more rapid introduction and, equally important, proper application of new technologies.

- Data collection and dissemination—collection and wide dissemination of data on both energy end use and technologies for energy efficiency, including information to the architects and engineers who design buildings.

Long Term

Table 4-9 shows that in the long term energy conversion technologies diminish in impact compared to building components and systems or design/practice. This is because the demand for space conditioning, especially heating, will be greatly reduced by building design and operation.

Advanced technology in all areas of energy use not affected by building design will make important contributions. Sustained basic and applied research is needed, beginning in the near term, on a wide variety of topics.

Energy Conversion Technologies. While existing energy conversion cycles offer major efficiency improvements, further opportunities may merit research programs in areas such as

- Sorption heat pump, cooling, and refrigeration cycles—absorption and adsorption cycles thermally driven by solar energy or a combination of gas (natural or biomass-derived methane or hydrogen) and solar energy.

- Solar photovoltaic-especially systems that can be integrated with a building's design, including photovoltaic coatings for windows.

- Advanced cogeneration systems—advanced heat engine systems and solid oxide fuel cells.

Building Components and Systems. R&D on building components and systems promises major impact on reducing GHG emissions from the buildings sector in the long term. Key research areas are as follows:

- Superinsulation—building walls, windows, and roofs as well as refrigeration systems, hot water storage, and other energy applications would all benefit from R&D on advanced insulation materials; non-CFC and non-GHG foams and evacuated panels are also important research avenues.

- Lighting—lighting requirements, optimal lighting wavelengths, and efficient lighting devices.

- Electronics—advanced controls and electronic systems for building energy management; high-efficiency electronics for reduced energy use for office equipment with the secondary benefit of reduced cooling loads.

- Construction materials—materials for building structures, finishes, and furnishings with less embodied energy and high carbon content; high growth rate structural timber; other biological materials.

Design/Practice. New technology for building design, commissioning, operation, and maintenance as well as low-energy, low-GHG emission concepts for community design can have significant impact on the buildings sector by the year 2050. Important research areas are as follows:

- Effects of the built environment on human behavior, health, and safety—spatial configurations and environmental conditions in buildings have profound effects on human performance, productivity, comfort, morale, health, and safety; interdisciplinary research is needed to better understand and predict these effects.

- Diagnostic technologies for buildings—appropriate research paths include nondestructive diagnostics to determine conditions of the building envelope and environmental control equipment of existing buildings; accurate and real-time diagnostic tools for quality assurance and condition assessment at construction sites; acoustic, electromagnetic, or other diagnostic concepts; establishing or verifying standards for ventilation, lighting, and air quality.

- Community design—urban and community planning techniques based on an improved understanding of the effects on human behavior from the above-recommended research.

Assessment of the Federal R&D Program

From FY 1974 (when the first legislated mandate for federally supported building energy R&D was enacted) to FY 1977, the conservation R&D budget grew at a very rapid rate from $12.8 million (in 1975 dollars) to $104.4 million (in 1977 dollars). In the same time period, the buildings budget (at the Energy Research and Development Administration, the predecessor agency to DOE) grew from $2.4 million to $35.6 million. With the establishment of DOE in 1977, the buildings budget increased modestly, given the inflation rates of this period, from $63.4 million in FY 1978 to $91.3 million in FY 1981. These programs led to the commercial introduction of such products as the heat pump water heater, solid-state ballasts, compact fluorescent bulbs, reflective window films, and advanced motor compressors. DOE also funded a vigorous solar energy research program that affected the building sector.

In 1981, at the urging of the new Reagan administration, Congress reduced DOE's FY 1981 buildings appropriation to $64.2 million and in FY 1982 appropriated $42.8 million. Since 1982 the buildings appropriation budget has been about $35 million. Throughout those years the administration has requested about $15 million annually for a few long-term, high-risk research programs, but Congress has added funds to enable a minimum critical level of activity and has initiated new programs such as least-cost utility planning.

During this time the administration requested to zero out the Center for Building Technologies (CBT) program at the National Institute for Standards and Technology, but Congress has continued to fund it at about $3.9 million annually for nonenergy programs. DOE has continued to fund CBT for energy conservation at a reduced level of about $1.9 million annually; DOE has eliminated support of solar energy research at CBT.

For FY 1991 the DOE buildings budget will likely be close to the FY 1990 appropriation of $36.8 million and activities will include near-term technology and prototype development. CBT will also likely be funded at its current level.

Despite some important results from past R&D on buildings, the current activities are not adequate to address all of the priority R&D activities cited above. For example, the FY 1990 DOE budget for the Buildings and Community Systems programs of $36.8 million is a minuscule percentage of the roughly $400 billion spent annually for construction. Also, since the major near-term impacts will come from implementation of existing technologies, DOE's budget should include funds for implementation, R&D.

One element of implementation R&D could include selected demonstrations although past demonstration activities have not been particularly successful. The committee recommends that DOE conduct a review of federally funded demonstration efforts in energy conservation in order to identify the factors that made some succeed and others fail. In conducting such an evaluation the definition of success must be based on whether the demonstration led to significant market adoption of the relevant technologies.

The committee also notes that the federal government itself constructs and operates a large number of buildings and is a major energy consumer. The total energy bill for the federal government is $8.5 billion annually, of which $3.7 billion is for buildings. For building-related energy conservation measures across the entire federal spectrum, $51.7 million is allocated or 1.4 percent of the federal building energy bill. To help reduce its energy use, the government has established a coordinating program called the Federal Energy Management Program. This program is funded at an annual level of $1.2 million.[11]

Government buildings are an obvious target for an implementation of energy-saving technologies but are not currently used for this purpose. The national laboratories, which have developed many energy-saving technologies, have tried but have not succeeded in securing funds to implement any of their technologies at their own government-owned facilities.[35]

The committee believes that a review of DOE's Building and Community Systems program is appropriate with participation from the National Institute of Standards and Technology, U.S. Department of Housing and Urban Development, General Services Administration, Gas Research Institute, Electric Power Research Institute, national laboratories, external industry, and university and public interest groups to set the nation's R&D agenda for the 1990s.

Technology-Adoption Strategies

Although many past energy technology development efforts cannot be labeled failures, market acceptance of efficient equipment has been slow. Analysis of these technologies continues to show that they offer clear life-cycle cost savings and reasonable payback periods to purchasers. Some of these problems can be offset by providing increased and accurate information. Several studies (see particularly reference 43 and its cited literature) have attempted to analyze consumer behavior in terms of investment for increased energy efficiency. A major finding[43] is that payback periods for investments in increasing energy efficiency of most household appliances are 2 years or less. Such studies conclude that the market for energy efficiency is not performing well.

The factors responsible for this situation are more pronounced in the buildings sector than in any other sector. Such factors include the following:

- Complexity of the decision-making structure, where buildings are often neither designed nor purchased by the ultimate tenants, and decisions are made instead by the builder, architect, construction firm, or financial institution.

- Difficulty of providing accurate information and feedback to millions of decision makers.

- Lack of education/training of building operations and maintenance personnel.

- Lack of energy-saving incentives for utilities or other institutions that are in a position to centralize decision making and provide access to the necessary capital.

For these reasons the buildings sector must rely to a greater extent on demonstration projects and national user facilities[44] to ensure that cost-effective technologies are actually adopted.

The following six actions must be considered to overcome these problems. These actions need to be followed for both the near- and long-term time frames to achieve the GHG emissions reduction potential of the buildings sector by ensuring the adoption of the technologies that result from the R&D programs recommended above.

Make Energy Service Markets Work

If energy were sold as a service rather than as a commodity, many problems could be overcome. The concept of least-cost energy service is not a new one,[45] but it has not been widely accepted. There are several approaches that might make energy service markets work. Two are considered to be the most likely to succeed:

- Pricing. One way to overcome the very high discount rates of many decision makers is to significantly increase the price of energy, thereby shortening payback periods. Increased attention to all energy use would also follow. Energy or environmental impact taxes are certainly one approach to increasing energy prices. "Revenue neutral" approaches also have been proposed and are worthy of consideration. Such approaches would place penalties on purchasers (or owners) of less energy efficient homes, buildings, equipment, etc., while providing for subsidies to purchasers of more efficient systems. The analogy is the "gas guzzler" tax but with the revenues being used to lower the price of vehicles providing high-mileage performance.

- Utility Regulation. Revised utility rate-making regulation could change the incentive structure for both gas and electric utilities. Reducing building energy consumption by 50 to 75 percent or greater reduces utility revenues almost proportionally in today's rate-making environment. The utilities are strategically placed to serve as unified decision makers in the buildings sector. Currently they do not have the incentives that would lead to reduced energy use and reduced GHG emissions. However, the committee notes that a few utility commissions in different areas of the country are beginning to address this problem.

Significant, and difficult-to-implement, changes are needed in utility regulation. For example, if a utility billed a residential customer for degree-hours of comfort provided (presumably based on a simple measurement of the indoor and outdoor temperatures), the utility could make cost-effective decisions and earn the same rate of return for its stockholders. The utility could make money and earn a return just as easily by deciding to add insulation, as by selling gas or electricity.

Provide Education and Training

The implications of energy use on the environment and the efficient use of energy must be taught. Effective publications, widely disseminated, can provide information to the user. Methods encouraging private sector marketers to include energy use information on all products ought to be pursued. Such methods might specify requirements for labeling energy-related products. For example, labeling windows and other building materials should be considered as well as energy-consuming devices.

Training programs for equipment installers, equipment operators, and building managers should be strongly encouraged. Such programs should exploit modern information and training technologies, such as interactive video disks and expert systems. As noted earlier in this report, improved information is needed on how energy is used in buildings, from construction through operation.

Improve Building Practices

The practices used in planning, designing, constructing, installing, operating, and maintaining buildings and the energy-providing and energy-using equipment need to be improved. This requires improved standards and methods that recognize the importance of energy and environmental concerns. Such standards need not always be mandated; model codes or standards are often broadly accepted.

In California builders have the option of using a set of prescribed formulas for construction or optimizing the home design based on a computer model. The optimized designs tend to be both more energy efficient and cheaper to build. Approximately two-thirds of the new homes in California are now designed this way.

In the Pacific Northwest, model codes developed by the Northwest Power Planning Council have been adopted as the de facto standard for construction.

The potential international aspects of quality standards should also be considered. Building materials and practices have not generally been a part of international commodity transactions. A set of internationally agreed-to practices for building construction that minimize GHG emissions would serve to expand the markets and encourage increased private sector investment.

Accepted measurement and diagnostic practices applied during and after construction of buildings are also needed to help meet GHG emission reduction goals.

Provide Feedback and Motivation

Proper operation and maintenance of energy-using equipment, control systems, lighting systems, etc., are important. Their effectiveness depends on the successful development and implementation of expert systems for diagnosis and feedback to building managers and operators. Such systems are expected to have major impacts only in the long term. Motivation of managers of multifamily residential buildings is equally important.[46]

Develop Urban Design Practices

Urban design practices could have a major impact on energy use, far beyond the building sector alone. Federal, state, and local policies should be developed with greater attention to energy and environmental implications. Many of the interrelationships between factors such as density, vehicle transportation, and use of mass transit are not yet well enough understood to establish such policies. Research on these issues is needed to establish effective policies.

Apply Foreign Technology

A near-term implementation plan may include the testing and further development of an existing foreign technology. For example, Japanese frost-free refrigerators that meet the DOE 1993 efficiency standards have been sold since 1986. Japanese refrigerators have been more efficient than U.S. models since 1980, but this fact was discovered by accident and only minimal monitoring and testing of foreign products have been done to date.

INDUSTRY

Energy Use and GHG Emissions

The industrial sector consumes 29 quads of energy per year, which is 36 percent of the total U.S. energy consumption. A breakdown of fossil fuel energy use and CO_2 emissions within key manufacturing industries is presented in Table 4-13 for 1985, the most recent year for which such data are available. About 40 percent of current total industrial energy use is for process heat; about 30 percent is for services provided primarily by electric power, notably machine drive; and about 20 percent is for chemical feedstock, construction asphalt, and metallurgical coal (which can be considered a feedstock). Industry is currently the source of about 35 percent of the total U.S. CO_2 emissions. Releases from inorganic carbonates used, for example, to produce glass and cement are included. Such miscellaneous sources of CO_2 are responsible for 5 to 10 percent of the industrial total, with combustion of fossil fuels responsible for over 90 percent.

Major Targets for Attention

The efficiency of energy use in industry can be improved and the form of energy used can be altered to reduce GHG emissions with existing technology, but both come at a cost. Industry chooses its uses and types of energy based on competitive economics. A reduction or switch cannot simply be mandated without repercussions. U.S. companies seek cost-effective locations throughout the world for their production and processing operations. Unilateral actions by the United States to stabilize GHGs must consider international industrial mobility, in order to avoid exporting industry and associated CO_2 emissions. Restated, U.S. technical capability to make rapid changes in processes and products is constrained by economic factors.

Industry is diverse. Furthermore, each industrial subsector differs from the others, and wide differences in processes exist even within a subsector. Thus, opportunities for reducing the emissions of GHGs have to be approached on multiple fronts. In addition, enormous changes have occurred within industry over the past 50 years, and changes will continue. Major shifts in the U.S. industrial base are just as likely, arising from developments in biotechnology, from recycling, and from material substitution, as well as shifts resulting from environmental considerations such as global warming and climate change. This should temper any rush to impose actions not justified by other relevant considerations.

Table 4-13 Fossil Fuel Use and Carbon Emissions by U.S. Manufacturing Industries, 1985[47]

	Natural Gas		Petroleum[b]		Coal	
	Energy Use[a] 10^{12} Btu	Emissions MTC	Energy Use[a] 10^{12} Btu	Emissions MTC	Energy Use[a] 10^{12} Btu	Emissions MTC
Petroleum	717	10	165	5	8	0
Chemicals	1680	24	855	17	332	8
Primary metals	693	10	60	1	1131	28
Paper	172	3	406	8	309	8
Stone, clay and glass	386	6	53	1	323	8

[a]Fuel use is expressed in 10^{12} Btu and carbon emissions in million metric tons (MTC). Conversion factors are expressed in million metric tons of carbon (MTC) per 10^{15} BTU of fuel: Natural gas, 14.5; petroleum, 20.3; coal, 25.1.

[b]Petroleum values are aggregates of residual fuel oil, distillate fuel oil, and liquid petroleum gas.

Options open to industry to reduce GHG emissions fall into four areas:

- improvement of energy efficiency,
- fuel switching,
- recycling of materials, and
- use of biomass-derived fuels and feedstocks.

Adoption of these options is subject both to the availability of technology and to economic factors. Relevant R&D opportunities and technology implementation issues are discussed in the sections that follow. Biomass for fuels and feedstocks represents a potentially new resource applicable to the other market sectors as well and is an option to be viewed over a longer time horizon. Since the production and use of biomass will require the establishment of a new industry infrastructure or substantial modification of existing industries, a brief perspective on this is provided in an addendum to this chapter.

Availability of Technology to Reduce GHG Emissions

Energy Efficiency Improvements

Energy efficiency as considered here involves the use of efficient equipment and improved operating procedures and production processes for reducing energy intensity, which is defined as energy use per unit of production. Energy efficiency improvements due to weight reductions per unit of product and material substitution in manufactured products are also important but are outside the scope of this study.

The baseline of industrial energy efficiency improvement can be placed in perspective by considering the period 1958-1971, when energy prices were low and even falling. In that period fossil fuel intensities declined about 1.2 percent per year as a result of ongoing technical changes. Electricity intensities increased about 1.8 percent per year, reflecting new applications of electricity.

In the period 1971-1985, fossil fuel intensities fell more rapidly, primarily in response to increases in fuel prices and secondarily in reaction to shortages of fuels and to government policy initiatives. Electricity intensities in the same period declined gradually, with gains from efficiency improvements outweighing demands for electricity in new applications.

Ongoing efficiency improvements will continue owing to the dynamics of industrial competition. The key question is: How can this baseline gain be accelerated?

Near Term. Changes in federal R&D are unlikely to have much effect on industrial energy efficiency in the near term. Nevertheless, a greater emphasis on coordination of generic process development by federal agencies and on the adoption of more efficient technology by industry could be of value. Examples of development areas are as follows:

- In the aluminum industry, use of electricity for smelting can be reduced from 7.0 kwh/lb to 6.0 kwh/lb.

- In manufacturing operations such as fabrication and assembly, improved use of electricity in lighting can save about 4 percent of electricity.

- Motor drives account for 67 percent of the electricity used by industry.[48] Adjustable-speed drives can save about 2 to 3 percent of total primary energy use attributed to the industrial sector.

Long Term. In several energy-intensive industry subsectors, federal support is needed, partly for basic research and partly to help structure long-term R&D programs. For example:

- In the steel industry, two major technologies in development are direct steel making and near-net shape casting. Compared to conventional processes, reductions in energy intensity and CO_2 emissions of more than 25 percent are projected when the technologies are adopted.

- In the aluminum industry, inert anode, cathode, and sidewalls could enable further reductions in the average intensity of smelting from 6 to 4 kwh/lb. Work on these important technologies is jointly funded by DOE and industry. These efforts should be sustained.

- In the pulp and paper industry, biodigestion, oxygen bleaching, dry forming of paper, and black-liquor gasification with combustion in gas turbines are technologies that could virtually eliminate the need for fossil fuels and purchased electricity.

- In the chemicals industry, biotechnology offers opportunities for efficient processing, and biomass feedstocks offer promise as substitutes for petroleum.

Potential Impacts. A substantial increase in energy efficiency will be achieved over the next 30 years through the adoption of currently available technology and the incorporation of results of the R&D efforts cited above (Table 4-14).

Fuel Switching

Fuel switching can lead to reduction of GHG emissions from selected industrial processes, but implementation depends on relative price (e.g., of electricity and natural gas). Switching fuels can be between different fuel types or between fuels and electricity. The first option is to switch from high-carbon fuels to fuels with lower carbon content (e.g., coal to natural gas). Switching to electrical energy is advantageous if the power is derived from sources such as hydro, solar, biomass, or nuclear, which are themselves not net producers of CO_2.

Generally, switching does not require new technology. The primary industrial use of fuels is for heating, and fuel switching is, in principle, a relatively straightforward operation. The problem is how to achieve it without imposing a major economic penalty. The prime focus for R&D ought to be on steps to lower the cost of natural gas and electricity. There is also need for continuing R&D on efficient electrotechnologies.

Table 4-14 Estimates of Energy Efficiency Potential by Industry, 1990-2020

Industry	Average Energy Intensity Reduction, %/year	Comments
Steel	1.0	Achievement of direct steel making in integrated mills and near-net shape casting with resulting 30 percent reduction in total energy requirement per ton of steel mill products.
Aluminum	0.5	Reduction from 7.0 to 4.5 kwh/lb of electricity requirements for smelting, assuming successful development of inert electrodes.
Chemicals	1.5	Source: D. Steinmeyer (member of the Industry Panel), Monsanto Co., personal communication, 1989.
Pulp and paper	2.2	A 49 percent reduction in total energy use per unit of production. (Source: DOE, Office of Industrial Programs, The U.S. Pulp and Paper Industry: An Energy Perspective, April 1988)
Glass	1.0	A 25 percent reduction in total energy use per unit of production. (Source: DOE, Office of Industrial Programs, The U.S. Glass Industry: An Energy Perspective, April 1988)
Fabrication and assembly	1.5	A 35 percent reduction in total energy use per unit of production. (Source: M. Ross [member of the Industry Panel], University of Michigan, personal communication, 1989).
Petroleum	0.5	Source: G. Lauer (member of the Industry Panel), Atlantic Richfield Co., personal communication, 1989.

Note: Estimates are based on the judgments of the Industry Panel members who participated in this study.

Near Term. The committee's overall strategy calls for the use of natural gas as an interim, low-CO_2-emitting fuel option in a variety of electric generation and end-use applications. Worldwide and U.S. domestic natural gas resources appear adequate to support its use in this role. However, the ability to supply and deliver the necessary fuel to end users within the appropriate time frame and price is a continuing concern. To address this concern, two general areas of R&D are appropriate.

First, R&D is needed to permit economic recovery of known domestic reserves. R&D should include

- advanced instrumentation to increase recovery of bypassed gas in existing producing or shut-in fields;

- evaluation of reserves and probable cost for recovery of deep gas (>15,000 ft);

- new methods to increase recovery of gas from tight formations;

- advanced technologies for small-scale gas separation to permit use of smaller fields of subquality (high N_2, H_2S, or CO_2 content) natural gas; and

- capture of biogenic methane from landfills and from municipal wastes.

These R&D topics are the subject of considerable effort by the oil and gas industry, including the Gas Research Institute.

Second, basic, longer-range research in geosciences should be pursued to examine other potentially economical gas resources to further extend the potential for natural gas as a transition fuel. Relevant studies should include resource evaluation of deposits of methane clathrates (hydrated methane) and recovery potential for methane dissolved in geopressured brines. This research should be directed at understanding the geology and size of the potential resource and the technology needed for future economic recovery.

Long Term. Hydrogen derived from electricity has the potential to be used in both the industrial and transportation sectors. In the steel industry, for example, technology can be developed to use hydrogen as the principal reductant in place of carbon. Future applications of the technology will, however, depend on the availability of very low cost electricity. The potential R&D focus should be to develop hydrogen as an alternate fuel to hydrocarbons with emphasis on such factors as safety, storage, and transport.

In the chemical, petroleum refining, and glass industries, electricity can, in principle, replace fuel for process heat. Such a replacement would, however, not be economical unless the price of electricity relative to that of fossil fuels were half or one-third the current ratio.

Potential Impacts. If it is assumed that all fossil fuels burned by manufacturers could be replaced by electricity and that all electricity is generated using nonfossil primary energy resources, about 40 percent of total current U.S. CO_2 emissions would be eliminated. (As noted earlier, this requires a major change in the energy resources for generating electricity and reductions in its cost to industry relative to competing fuels.)

Recycling of Materials

Increased recycling offers a near-term opportunity for reduction of energy and CO_2 emissions in industry. The important subsectors are primary metals, pulp and paper, organic chemicals, petroleum refining, and glass.

Major issues associated with the potential for achieving recycling are

- creation of markets for postconsumer-recycled material in the manufacture of high-quality products;

- mechanisms for reliable and clean collection of selected postconsumer and industrial waste materials; and

- regulatory changes to allow currently defined "waste" streams to be used as feedstocks, both within a single industry and between industries.

Proper technology could significantly enhance effective end-user (consumer) separation of recyclable resources. In principle, the majority of domestic waste can be recycled into valuable feedstocks. Aluminum beverage cans, made from about 50 percent recycled aluminum, are an outstanding example of a technologically very demanding product for which a system for the supply and use of recycled material has been developed.

With respect to recycling within industry, changes are needed in regulations such as the Resource Conservation and Recovery Act of 1976, which strongly discourages the utilization of waste streams across corporate boundaries.

Near Term. The following actions can be taken in the near term to promote recycling:

- Design products so that their recycling can be achieved more efficiently.

- Develop separations technology to provide pure materials that can be used to produce high-quality consumer and industrial products.

- Develop new or improved processes to economically utilize a larger fraction of industrial waste materials.

- Establish standards for marking or otherwise labeling containers and container material to facilitate efficient separation. The standards should be such that physical classification and separation equipment can be built to handle the enormous amount of material currently being deposited in landfills.

In the area of consumer waste materials (such as cans, bottles, plastics, and paper), the technology to separate wastes is available but is not widely implemented. In the area of industrial waste, such as bauxite residue (red mud) and residue from copper production, separation technology needs to be developed. A specific example of needed technological development is separation of plastics, where there are 15 major varieties with which to contend. In this instance the technological development could be aimed at the incorporation of "label" molecules into all plastics; this label could be read at a separation facility and permit easy separation.

Potential Impacts.

- The recycling rate of aluminum is currently 55 percent. Increasing this rate to 75 percent would save approximately 0.3 quads/year of energy and reduce the attendant CO_2 emissions.

- With maximum effort, 50 percent recycling of plastics might be achieved by the year 2000. The corresponding reduction in energy used by the chemical industry would be about 15 percent.

- An increase in the quantity of recycled glass containers reduces CO_2 emissions by 5 percent for each 10 percent of recycled glass. This is achieved through reductions in both energy requirements and in the quantity of carbonate raw materials contained in the feedstock.

- The iron and steel industry has always recycled nearly 50 percent of its output (nearly 40 percent of domestic active capacity is based entirely on the sole use of scrap). Two areas for increased recycling are the recovery of coated scrap (e.g., scrap steel "cans" from the after-use market) and the mixed grades of scrap from automobile hulks and demolitions.

- Almost 20 million short tons of waste paper and paperboard were reused in 1987, representing 26 percent of paper and paperboard production. Some 4.5 million tons of waste paper was exported. Almost no coated or glossy papers are recycled now because the present coatings are incompatible with reuse.

Biomass

Biomass offers an opportunity for reducing CO_2 emissions because it has the potential to supply process energy, feedstock for chemicals, and transportation fuels on a CO_2-neutral basis.

Near-Term.

- Continue and enhance studies on

  improved conversion of biomass, especially wood derived cellulose and hemicellulose, to ethanol and other products;

  improved anaerobic digestion of farm and municipal wastes to produce methane; and

  selection of species that provide high yields of dry biomass and have low energy requirements during processing.

Long Term.

- Continue and expand understanding of

  the mechanisms of photosynthesis;

  genetic factors that influence plant growth and those that enhance high yields of desirable plant properties;

  symbiotic process between certain bacteria and plants that permits the fixation of nitrogen (legumes);

  genetic factors that enhance plant tolerance to the environmental stresses of drought, excessive heat/cold, and disease; and

  plant biodiversity and habitat implications of biomass.

- Develop plants incorporating the knowledge gained from the steps outlined above.

Potential Impacts.

- In principle, a CO_2-neutral economy based on biomass is possible. It is a very long range goal, and research is needed to determine if it is achievable.

- Estimates for the potential energy contribution from biomass extend to 26 quads of dry biomass/year.[49] The 26 quads/year reflects optimistic assumptions of land availability and is based on yields achieved on carefully nurtured, relatively small plots.*

- Although 26 quads/year is much higher than current biomass contributions (approximately 3 quads/year), it reflects only modest contributions from recombinant DNA (biotechnology). If this developing science follows the cycle of invention/innovation/contribution of past scientific breakthroughs, major contributions could be anticipated starting in the first quarter of the twenty-first century.

- At a minimum,

  biotechnology could reduce the need for chemical fertilizer nitrogen supply. Almost 2 percent of the industrial CO_2 output comes from production of nitrogen fertilizer.

- At a maximum,

  the chemical industry would switch much of its feedstock to biomass, the paper and forest products industries would become smaller uses of agricultural land and greater exporters of by-product energy, and the petroleum industry would switch to a biomass base for methane and liquid fuels production.

----

*From surplus cropland: $(103 \times 10^6 \text{ acres}) \times (152 \times 10^6 \text{ Btu/acre/year}) = 15.7 \times 10^{15}$ Btu/year.

From marginal land, not currently cropped: $(89 \times 10^6 \text{ acres}) \times (116 \times 10^6 \text{ Btu/acre/year}) = 10.3 \times 10^{15}$ Btu/year.

R&D Needs and Priorities

A multiple path approach should be followed to achieve a reduction of GHG emissions by industry. Energy efficiency improvements and recycling could offer major benefits in the short term and will facilitate the switch to other fuels and biomass in the future. Substantial R&D is needed to verify the biomass option, and electrical generating capacity using non-fossil fuels cannot be readily and cost-effectively expanded—hence, electrification and biomass are long-term options.

Energy Efficiency Improvements

- Continued energy-efficiency improvements offer a major opportunity for reduction in CO_2 emissions. In only a few production processes do current efficiencies approach thermodynamic limits.

- Changes in federal R&D would have little effect on industrial energy efficiency gains in the near term.

- In the long run, federal support of basic and generic research would be of value to the energy-intensive industry subsectors.

Moderate gains in efficiency will continue to be made and such improvements will occur without new policies. A significant—but not major—acceleration of the improvement rate can be achieved through new policies targeted at reducing the costs of energy-conserving equipment.

Federally sponsored R&D must have industry guidance and feedback as well as clear objectives. The Metals Initiative exemplifies a promising approach, wherein DOE's objectives include gains in both energy efficiency and U.S. industrial competitiveness.

In addition, government assistance in R&D can be provided through user-oriented research centers (such as those funded by the National Science Foundation, the National Institute of Standards and Technology, and the U.S. Department of Defense) for particular generic manufacturing-process issues, like adhesives and joining, forming of ceramics, standardized characterization of plastics, sensors and process controls, and combustion.

Fuel Switching

With regard to fuel switching, government policy actions must be guided by the following considerations relative to current practice:

• Costs must be lowered and capacities increased for electricity generation using low- or non-GHG-emitting resources, large reserves of methane, and low-cost biomass-derived fuels.

• Availability of very low cost electricity would be essential for developing hydrogen as an energy resource.

A variety of actions could be taken to expand methane supplies for industrial use. For example, gas producers such as those operating on municipal landfills could be permitted to use the existing pipeline infrastructure. In addition, international agreements could be negotiated for the recovery and use of gas that is flared and vented.

Recycling of Materials

Increased recycling offers a major opportunity for increasing energy efficiency and reducing CO_2 emissions. The important subsectors are primary metals, pulp and paper, organic chemicals, petroleum refining, and glass. To stimulate recycling the following issues must be addressed:

• Front-end separation efforts may have to be encouraged through incentives or penalties.

• Markets must be created for postconsumer-recycled material in the manufacture of high-quality products.

• Mechanisms are needed for reliable and clean collection of selected postconsumer and industrial waste materials.

• Regulations counterproductive to waste management initiatives should be changed. For example, regulations must be changed to allow the use of currently defined "waste" streams as feedstocks, both within a single company and between companies.

• Industries ought to be established that use the waste of other industries to produce useful products cost-effectively.

The structure and substance of federal, state, and local regulations affecting waste management are, in some instances, counterproductive. Implementation and interpretation of the current law have strongly discouraged the recycling of materials that have historically been designated waste and have not been sold in commerce. The principal problem is a matter of liability for the ultimate disposal of the hazardous waste. The statutory language should be amended in such a manner that it encourages the utilization of waste streams and materials across all industries.

Although technologies—ranging from magnetic and electrostatic sorting to plasma decomposition and separation of the elements by mass spectrometry—could be developed for separation of mixtures, the most cost-effective and simplest method for nonindustrial wastes is front-end sorting at the household. Here the separated plastics, metals, paper, and glass can be shipped to the proper industry for purification and recycled into products having high value added. Since front-end separation requires extra effort on the part of the consumer or scrap-metal dealer, this effort may have to be encouraged through incentives or penalties imposed through legislation.

In the area of industrial waste, there is high potential for the establishment of industries that use the waste of other industries to produce useful products, without the need for mining and comminution, both of which use substantial energy. An example is metals recovery from electric-furnace dust. Federal leadership and financial support will be required if U.S. industry, academia, and the national laboratories are to launch a broad R&D effort to achieve the maximum amount of industrial recycling.

Biomass

To ascertain the full potential of the biomass option, a greatly increased, coordinated, and focused federal program is needed. The areas for federal focus in R&D are to

- expand understanding of basic plant science,
- support short-term actions that focus on conventional farms, and
- perform systems analysis to define and prioritize infrastructure requirements.

At the same time, industry should pursue R&D to

- develop more efficient plant species and
- develop plant species that can adapt to specific end uses (i.e., pest, nutrient, and climatic environments).

Areas where broad, federally chartered, cost-shared, industrially guided R&D programs should be considered include

- development of biomass plantation concept, and
- development of conversion processes.

In addition, the scientific base to regulate introduction of genetically altered species needs to be strengthened. There are concerns about wide-scale use of biotechnology that need to be addressed by means of much improved scientific knowledge—for example, the potential loss of genetic diversity.

Incentives are needed to manage farms for long-term efficiency, through erosion control and more efficient irrigation, and to operate farms as integrated systems (e.g., manure utilization). Incentives ought to be maintained to nurture the emerging use of biomass for energy, feedstock, and liquid fuels (e.g., ethanol in gasoline).

Mechanisms must be established to broadly recognize, assess, and manage societal and environmental impacts pertaining to the biomass industry, including use of water resources, competition between food and fuel/feedstock for land use, diversity of plant species for protection against pest attack, and preservation of biodiversity and habitat.

Recognizing the need for protection of the environment is a major issue for large-scale development of biomass as an energy source.

Current Energy R&D Programs

The federal government has a broad spectrum of R&D programs supporting energy conservation, the development of renewable energy sources, and low-cost electricity production. These three areas are central to industry's opportunity to reduce GHG emissions. Assessing the effectiveness of these federal R&D programs should take into account that the results of basic research are distant in time; frequently, the spark that led to invention, innovation, and application is forgotten at the time of application. Federal funding also is relatively slight when seen as a percentage of the total worldwide historical research base that underlies many industries.

Historically, DOE has concentrated its research and development funding on supply technologies, not on end-use technologies. Energy R&D within DOE that is relevant to the requirements of industry is mainly within the Conservation and Renewable Energy program areas. The relative program priorities within DOE, however, as measured by the funding assigned to the Conservation and Renewable Energy program areas in FY 1989 were among the lowest (see Table 3-1). Aside from its small scale, the federal effort on conservation and renewables may have a problem in that it is distributed too broadly and lacks a focused thrust. Two areas in which the committee recommends focus are recycling (for contribution in the period 1990-2010) and biomass (for contribution in the period 2000-2030).

DOE's Office of Industrial Programs (OIP), which in FY 1990 is funded at $51 million, addresses many key industrial R&D needs to achieve more efficient energy use.[11] In addition to the federal funds spent by OIP, industry will spend about $8 million in FY 1990 in support of these cost-sharing projects.

The planned activities of OIP in FY 1990 include a budget of $27 million for R&D on improved energy productivity. About half of this funding is for advanced steel-processing research projects jointly funded by industry. Funding for other energy productivity research projects is generally small and directed at many targets. Further review of these projects should be done in collaboration with industry to determine if OIP funds should be focused differently.

Recycling is being addressed in modestly funded projects on separation systems and industrial waste utilization. A longer and more comprehensive program in these two areas is required to develop technologies to make recycling a viable option.

The waste, heat recovery, combustion improvement, and industrial cogeneration projects are also modestly funded (a total of about $16 million in FY 1990) and are directed at generic technologies in support of industry. More detailed review by industry is desirable to determine if the scope and funding of these projects are appropriate.

The Biofuels and Municipal Waste Technology Division of DOE is assigned responsibility to develop fuel pathways for producing large quantities of domestic biomass and to convert it to quality fuels at minimum cost. Development of new high-yield energy crops and biotechnology are a part of this program area. The projects within this program have well-defined objectives and appear to be making substantial progress toward improving a number of biomass technologies. Total funding for the Biofuels Division in FY 1990 was about $16 million. This level of funding is inadequate to meet the identified technological needs for biomass production and conversion.

Within the basic research activities of DOE, the chemistry, materials, and biosciences areas provide a phenomenological basis upon which industry could develop or improve process technologies and products. Approximately $400 million will be spent by DOE in these areas in FY 1990. A closer tie between basic research at DOE and RD&D in the applied program offices would be valuable.

Other federal agencies also sponsor process technology R&D for industry applications. Such activities include manufacturing process technology development for the metals fabrication and electronics industries by the U.S. Department of Defense (FY 1990 funding of $170 million); manufacturing R&D, especially process control and automation, by the U.S. Department of Commerce through the National Institute for Standards and Technology; the development of improved materials technologies by the Bureau of Mines; biomass and other agricultural energy improvement techniques by the U.S. Department of Agriculture; and, at the very important basic research level, the science and engineering research programs of the National Science Foundation. Certain R&D activities of the U.S. Environmental Protection Agency and the U.S. Department of Transportation are also pertinent to industry. More coordination among these activities would be desirable to ensure that technology transfer occurs across federal agencies and to industry.

Although each of the federal R&D programs can directly and indirectly contribute to the base of knowledge needed by industry to address the GHG problem, none were formulated with this as a prime target. As noted earlier, the relative priorities will shift if the GHG issue becomes a major consideration in federal R&D, with biomass being one obvious beneficiary.

ADDENDUM: BIOMASS FOR ENERGY AND FEEDSTOCKS

All products derived from photosynthetic activity—agriculture, silviculture, and aquaculture—are encompassed by the term biomass. This resource offers a major nonelectrical route for achieving large reductions in CO_2 emissions by supplying process energy, feedstock for chemicals, and transportation fuels in a CO_2-neutral manner.

On a worldwide basis, biomass is already a significant source of energy. Although less than 1 percent of the annual biomass growth is used for energy, it provides 15 percent of total primary energy consumption.[50] In individual countries the biomass contribution is much higher: 25 percent in Brazil, 33 percent in China, and 50 percent in India. At present, biomass supplies less than 5 percent of the energy used in the United States. However, the United States is relatively rich in biomass generated, with approximately 50 quads of energy fixed as biomass each year and about half of that used as harvested product.[51] In terms of harvested product, that is about 2.4 times as much as that in Brazil and 5 times as much as that in India.[51]

Most of the U.S. harvest goes to food and pasture, with 7 percent going to forest products.[51] The energy content in the current harvest is equal to about a third of current energy use. If it were to be shifted into liquid fuels via the relatively inefficient corn-to-ethanol cycle, the fraction would be much smaller.

Biomass production for energy and feedstocks faces a number of constraints, including competition with food and fiber production, land and water shortages, and environmental degradation.[52] The potential damages from biomass development can involve substantial increases in soil erosion and sedimentation of rivers and lakes and subsequent damage to land and water resources, adverse changes in or loss of important ecosystems, degradation of esthetic and recreational values, and local air and water pollution problems.[53] Although all of these adverse effects can be minimized through proper management, an assessment must be made on a region-by-region basis to determine the full potential of the biomass resource and the benefits and risks of its expanded use for fuels and chemical feedstock.

The amount of land needed in the United States for biomass production is directly dependent on the amount of energy and feedstock demands. Some relevant statistics[54] are as follows:

Areas	Million Acres
United States—total	2,265
Commercial forest	715
Cropland	475
Pasture	740

A total of 192 million acres was utilized in the estimate for the production of 26 quads of energy from biomass.[49]

Under the current DOE biofuels program, a potential recovery of biomass-derived energy of 17.0 quads/year has been projected.[51] Such a production level in the United States assumes the use of the following raw resources:

Biomass Resources	Recoverable Energy (quads/year)	Currently Recovered (quads/year)[50]
Conventional wood and forest waste	7.5	2.7
Energy crops (wood and herbaceous)	4.0	
Agriculture	1.0	0.1
Oil-bearing plants (oilseed and algae)	2.0	
Municipal waste	2.5	0.1

The energy yield can be increased with current technology. It has been demonstrated [49,55-57] that biomass yields from high-intensity plantations can achieve 10 to 25 dry tons of biomass per acre per year using specially developed varieties of sorghum, napier grass, and wood grass. This represents up to an order of magnitude increase above current average agricultural yields. A by-product of plantation cropping of fast-growing forests is the carbon fixation both in the standing forests and in their root systems.

Perhaps more important is the expectation of genetic improvements in the next 10 to 40 years. Currently, biomass captures only 0.1 percent of the average incident solar energy, which suggests the enormous potential for improvement. The reason to expect realization of improvement is the fast development of biotechnology—the understanding of the genetic code and the ability to select and insert desired genes. "Today over two dozen species of crop plants can be routinely transformed..... Genetically engineered soybean, cotton, rice, corn, and alfalfa crops are expected to enter the marketplace between 1995-2000."[56]

Biomass can be burned directly as a source of fuel or converted into gas or liquid fuels.[58] In the United States, approximately 1 percent of automobile fuel is ethanol derived from corn. In Brazil a somewhat larger program derives ethanol from sugar cane. In both cases a large subsidy is required to permit ethanol to compete against conventional gasoline. Also in both cases, slightly over half the cost of production is in the corn or sugarcane.[51] Both technologies appear relatively mature. There is little promise of a major rise in efficiency or a reduction in production costs.

The sugar cane cycle represents a source of automotive fuel that is nearly CO_2 neutral. The corn cycle is not CO_2 neutral, because significant amounts of fossil fuel are used in producing the corn and distilling the ethanol. The corn cycle in the United States appears to be motivated primarily as an indirect subsidy of agriculture, with a secondary theme of reducing the balance of payments and dependence on imported crude oil.

Few people suggest the corn-to-ethanol route as a logical long-term solution to transportation fuels. Typical estimates of the maximum contribution from corn-ethanol are two to three times its current level.[59] Still, the route is important as a pioneering effort, to develop the infrastructure for more extensive production and use of biomass fuels—and for that reason the subsidy may be worthwhile.

Most of the current work at the Oak Ridge National Laboratory, Gas Research Institute, and Solar Energy Research Institute has focused on other sources, such as glucose from cellulose and xylose from hemicellulose. Considerable progress has been made in lowering costs for this route, increasing raw material yields, and improving conversion processes. For example, projected costs for obtaining ethanol from cellulose and hemicellulose have been halved in the last 10 years—to about $1.35 per gallon today.[60]

NOTES AND REFERENCES

1. Annual Energy Review, 1978, Report DOE/EIA-0384, U.S. Department of Energy, Energy Information Administration, Washington, D.C., May 1989.

2. Technical Assessment Guide, Vol. 1, Report P-4463-SR, Electric Supply, Electric Power Research Institute, Palo Alto, Calif., 1986.

3. D. Golomb, et al., Feasibility, Modeling, and Economics of Sequestering Power Plant CO_2 Emissions in Deep Ocean, Report MIT-EL-89-003, Massachusetts Institute of Technology, Cambridge, Mass., December, 1989.

4. K. Block, C. Henricks, and W. Turkenburg, The Role of Carbon Dioxide Removal in the Reduction of the Greenhouse Effect, IEA/OECD Expert Seminar on Energy Technologies for Reducing Emissions of Greenhouse Gases, Paris, France, April 13-14, 1989.

5. Hydroelectric Power Resources in the United States, Federal Energy Regulatory Commission, Washington, D.C., 1988.

6. R. H. Williams and E. D. Larson, T. B. Johansson, B. Bodlund, and R.H. Williams, (eds.), "Expanding Roles for Gas Turbines in Power Generation," in Electricity: Efficient End-Use and New Generation Technologies, and Their Planning Implications, Lund University Press, Lund, Sweden, 1989.

7. Meridian Corporation, Characterization of U.S. Energy Resources and Reserves, DOE Contract DE-AC01-86CE30844, June 1989, p. A-29,.

8. Carl J. Weinberg, Pacific Gas and Electric Company, San Ramon, Calif., personal communication, 1990.

9. D. Carlson (Vice President and General Manager of the Thin-Film Division, Solarex, Newtown, Pa.), "Low-Cost Power from Thin-Film Photovoltaics," T. B. Johansson, B. Bodlund, and R. H. Williams, (eds., in Electricity: Efficient End-Use and New Generation Technologies and Their Planning Implications, Lund University Press, Lund, Sweden, 1989.

10. K. Zweibel and H. S. Ullal, "Thin-Film Photovoltaics," paper prepared for the 24th Intersociety Energy Conversion Engineering Conference, Washington, D.C., August 6-11, 1989.

11. Fiscal Year 1991 Congressional Budget Request, Report DOE/MA-0398 U.S. Department of Energy, Washington, D.C. , January 1990.

12. D. J. McGroff, presentation to the National Research Council Committee on Alternative Energy Research and Development Strategies, October 5-7, 1989.

13. United States Energy Policy, 1980-1988, Report DOE/S-0068, U.S. Department of Energy, Washington, D.C., October 1988.

14 J. D. Griffith, presentations to the National Research Council Committee on Future Nuclear Power Development, October 18 and November 13, 1989.

15. National Research Council, Pacing the U.S. Magnetic Fusion Program, National Academy Press, Washington, D.C., June 1989.

16. S. C. Davis, D. B. Shonka, et al., Transportation Energy Data Book: Edition 10, Report ORNL-6565, Oak Ridge National Laboratory, Oak Ridge, Tenn., September 1989.

17. D. L. Bleviss, The New Oil Crisis and Fuel Economy Technologies—Preparing the Light Transportation Industry for the 1990s, Quorum Books, New York, 1988.

18. M. Ross, "Energy and Transportation in the United States" Annual Review of Energy, 14: 131-171, J. M. Hollander, R. H. Socolow, and D. Sternlight, D. (eds.), 1989.

19. Holtberg, P. D., et al., 1988 GRI Baseline Projection and U.S. Energy Supply and Demand to 2010, Strategic Analysis and Energy Forecasting Division, Gas Research Institute, Chicago, Ill., 1988.

20. Nonresidential Buildings Energy Consumption Survey: Characteristics of Commercial Buildings 1986, DOE/EIA-0246 (86), U.S. Department of Energy, Energy Information Administration, Washington, D.C., 1988.

21. J. Bluestein and H. DeLima, Regional Characteristics and Heating/Cooling Requirements for Single-Family Detached Houses, GRI-85/0164, Applied Management Sciences, Inc., for Gas Research Institute, August 1985.

22. W. Q. Zwack, et al., Review and Comparison of GRI Single-Family Detached House Heating and Cooling Loads, GRI-86/0163, Applied Management Sciences, Inc., for Gas Research Institute, Chicago, Ill., December 1986.

23. E. Hirst, Cooperation and Community Conservation, Comprehensive Report, Hood River Conservation Project, Contract DE-AC-79-83BP11287, U.S. Department of Energy, Washington, D.C., 1987.

24. W. D. Houle, "Control System Usability," Strategies for Reducing Natural Gas, Electric and Oil Costs, In Proceedings of the 12th World Energy Engineering Congress, Atlanta, Ga., 1989.

25. R. Anderson and T. Hartman, "Controls of the Future," Heating/Piping/Air Conditioning, November 1988, p.59-61.

26. D. P. Fioriono, "An Application of State-of-the-Art HVAC and Building Systems," Energy Eng. 85(6):6-31, 1988.

27. V. E. Gilmore, "Superwindows," Popular Sci., March 1986.

28. T. Miyairi, "Introduction to Small Gas Engine-Driven Heat Pumps in Japan—History and Marketing," ASHRAE Trans. Vol. 95, Part 1, 1989.

29. C. E. French, F. E. Jacob, T. A. Klausing, and T. R. Roose, "Reciprocating Natural Gas-Engine Vapor-Compression Heat Pump," in Proceedings of the 1989 International Gas Research Conference, Vol. II: Residential & Commercial Utilization, Tokyo, Japan, November 6-9, 1989.

30. American Public Power Association, "Air-Source Heat Pumps Evolve," Air Conditioning, Heating, and Refrigeration News, October 1989, p.12.

31. D. S. Teji, "HVAC Equipment Replacement Study—Energy Savings Three Ways," 12th World Energy Engineering Congress (WEEC) Product Showcase, Atlanta, Ga., 1989.

32. Sylvania Lamps, An Energy-Saving Guide for All Your Lighting Needs, GTE Products Corporation, Sylvania Lighting Center, 1989/90.

33. R. R. Verderber, "Advanced Lighting Technologies Products," Strategies for Reducing Natural Gas, Electric and Oil Costs, in Proceedings of the 12th World Energy Engineering Congress, Atlanta, Ga., 1989.

34. D. Goldstein, Deriving Power Budgets for Energy-Efficient Lighting in Non-residential Buildings, American Council for an Energy-Efficient Economy, Summer Study on Energy Efficiency in Buildings, Washington, D.C., 1988.

35. Lawrence Berkeley Laboratory, presentation to the Buildings Sector Panel, National Research Council Committee on Alternative Energy Research and Development Strategies, November 14-15, 1989.

36. K. G. Davidson, "Advances in HVAC Alternatives". Heating/Piping/Air Conditioning, September 1987, p. 59-68.

37. J. R. Watt and A. A. Lincoln, "Refrigeration Systems Enhancement Thru Evaporative Cooling," Strategies for Reducing Natural Gas, Electric and Oil Costs, in Proceedings of the 12th World Energy Engineering Congress, Atlanta, Ga., 1989.

38. 1990-1994 Research and Development Plan and 1990 Research and Development Program, Gas Research Institute, Chicago, Ill., 1989.

39. Results of Appraisal of GRI 1990-1994 R&D, Gas Research Institute, Chicago, Ill., 1989.

40. If increased natural gas use is accompanied by significant (5 percent) leakage, the benefits of substituting natural gas for other fuels will be lost, since methane, the primary constituent of natural gas, is also a greenhouse gas. However, recent studies show leakage on the order of 1 percent or less (W. M. Burnett, Gas Research Institute, personal communication, 1990).

41. Household Energy Consumption and Expenditures. Part 1: National Data, DOE/EIA-0321/1, U.S. Department of Energy, Energy Information Administration, Office of Energy Markets and End Use, Washington, D.C., 1987.

42. P. J. Camejo and D. C. Hittle, "An Expert System for the Design of Heating, Ventilating, and Air-Conditioning Systems", ASHRAE Trans., Vol. 95, Part 1, 1989.

43. H. Ruderman, M. D. Levine, and J. McMahon, "The Behavior of the Market for Energy Efficiency in Residential Appliances Including heating and Cooling Equipment," Energy Jour., 8(1):101-123, 1987.

44. National user facilities are physical locations at government laboratories where potential users of new technologies, including industry manufacturers, professional associations, and other interested groups, can take advantage of a facility's staff and services on an as-available basis. The facilities conduct primarily nonproprietary testing and disseminate results widely.

They serve as R&D facilities where future advancements are pursued in parallel with implementation of existing technologies. Technology areas appropriate for user facilities include windows, roofing, lighting, construction materials, and operation and maintenance practices. User facilities must be established in close cooperation with appropriate trade organizations and must place high priority on availability to user groups. National user facilities exist at Lawrence Berkeley Laboratory for windows and Oak Ridge National Laboratory for roofing. The National Institute of Standards and Technology provides a user facility in the areas of thermal performance of walls; plumbing and water heating; appliance efficiency; commissioning, operating, and maintenance procedures for energy management and control systems; and durability of construction materials. Seattle City Light operates a user facility for regional users on lighting.

45. R. Sant and S. Carhart, Eight Great Energy Myths: The Least-Cost Energy Strategy—1978-2000. Energy Productivity Report No. 4, Mellon Institute, Pittsburgh, Pa., 1981.

46. R. Diamond, and P. du Pont, "Building Managers: The Actors Behind the Scene," Home Energy, March/April 1988.

47. Manufacturing Energy Consumption Survey: Consumption of Energy in 1985, Report DOE/EIA-0512(85), U.S. Department of Energy, Energy Information Administration, Washington, D.C., November, 1988.

48. L. Lamarre, "New Push for Energy Efficiency," EPRI J., 15(3):4-17, 1990.

49. J. Ranney, (Oak Ridge National Laboratory), presentation to Industry Panel, National Research Council Committee on Alternative Energy Research and Development Strategies, November 30, 1989.

50. D. L. Klass, "The U.S. Biofuels Industry," International Renewable Energy Conference, Honolulu, Hawaii, September 18, 1988.

51. D. Pimental et al., "Food Versus Biomass Fuel: Socioeconomic and Environmental Impacts in the United States, Brazil, India and Kenya," Adv. Food Res. 32: 185, 1988.

52. Report on Biomass Energy, Energy Research Advisory Board, U.S. Department of Energy, Washington, D.C., 1981.

53. Energy from Biological Processes, Office of Technology Assessment, U. S. Congress, Washington, D.C., July 1980.

54. S. R. Bull (Solar Energy Research Institute), presentation to Industry Panel, National Research Council Committee on Alternative Energy Research and Development Strategies, November 9, 1989.

55. T. D. Hayes, (Gas Research Institute), presentation to Industry Panel, National Research Council Committee on Alternative Energy Research and Development Strategies, November 9, 1989.

56. R. T. Fraley, "Genetic Engineering in Crop Agriculture," background paper for Office of Technology Assessment, U. S. Congress, Washington, D.C., October 10, 1989.

57. D. L. Kulp (Ford Motor Co.), presentation to Industry Panel, National Research Council Committee on Alternative Energy Research and Development Strategies, November 9, 1989.

58. National Materials Advisory Board, National Research Council, Bioprocessing for the Energy-Efficient Production of Chemicals, National Academy Press, Washington, D.C., April 1986.

59. Ethanol and Policy Tradeoffs, U.S. Department of Agriculture, Washington, D.C., January 1988.

60. R. J. Van Hook, presentation to Industry Panel, National Research Council Committee on Alternative Energy Research and Development Strategies, November 9, 1989.

BIBLIOGRAPHY

In addition to the references cited, the following sources were used in the preparation of chapter 4.

American Solar Energy Society, Assessment of Solar Energy Technologies, Boulder, Colo., 1989.

Andrews S., "New Developments in Integrated HVAC," Builder, pp. 114, 120, May 1988.

Brickman, P. E., "Commisioning: Why We Need It-What Are the Benefits?" ASHRAE Trans., Vol. 95, Part 1, 1989.

Brodrick, J. R., Commercial Buildings, Energy Consumption, and Natural Gas Markets, Gas Research Institute, Chicago, Ill., June 1986.

Brodrick, J. R., and R. F. Patel, "Assessments of Gas-Fired Cooling Technologies for the Commercial Sector," ASHRAE Trans., Vol. 95, Part 1, 1989.

Burnett, W. M. and S. D. Ban, "Changing Prospects for Natural Gas in the United States," Science, 244: 305-310, April 1989.

Committee on Innovative Concepts and Approaches to Energy Conservation, Energy Engineering Board, Commission on Engineering and Technical Systems, National Research Council, Innovative Research and Development Opportunities for Energy Efficiency, National Academy Press, Washington, D.C., 1986.

DeLuchi, M. A., et al., "Methanol vs. Natural Gas: A Comparison of Resource Supply, Performance, Emissions, Fuel Storage, Safety, Costs, and Transitions," SAE Paper 881656, Society of Automotive Engineers, Warrendale, Pa., 1988.

Friedlander, G. D. "Smart Structures," Mech. Eng., 110: 78-81, October, 1988.

Gopal, R., "An Evaluation Process for EMCS. Strategies for Reducing Natural Gas, Electric and Oil Costs," in Proceedings of the 12th World Energy Engineering Congress, Atlanta, Ga., 1989.

Hatfield, J. R., and B. B. Lindsay, "Development of a Gas Engine Driven Rooftop Air Conditioning Unit for the Commercial Market, 1989," Strategies for Reducing Natural Gas, Electric and Oil Costs, in Proceedings of the 12th World Energy Engineering Congress, Atlanta, Ga., 1989.

Houghton, R. A., and Woodwell, G. M., "Global Climate Change," Sci. Am., 260(4):36-44, April 1989.

Johnson, C. A., R. W. Besant, and G. J. Schoenau, "Economic Analysis of Daylit and Non-daylit Large Office Buildings for Different Climatic Locations and Utility Rate Structures," ASHRAE Trans., Vol. 95, Part 1, 1989.

Johnson, C. A., R. W. Besant, and G. J. Schoenau, "Economic Parametric Analysis of the Thermal Design of a Large Office Building Under Different Climatic Zones and Different Billing Schedules," ASHRAE Trans., Vo. 95, Part 1, 1989.

Klems, J. H., "U-Valves, Solar Heat Gain, and Thermal Performance: Recent Studies Using the MoWiTT," ASHRAE Trans., Vol. 95, Part 1, 1989.

Kurosawa, S. Y., et al., "Development of a High-Efficiency, Small-to-Medium-Sized Gas Absorption Chiller/Heater," in Proceedings of the 1989 International Gas Research Conference, Vol. II: Residential & Commercial Utilization, Tokyo, Japan, November 6-9, 1989.

Mei, V. C., and E. A. Nephew, Life-Cycle Cost Analysis of Residential Heat Pumps and Alternative HVAC Systems, Oak Ridge National Laboratory, ORNL/TM-10449, Report Brief, Oak Ridge, Tenn., 1989.

National Research Council, Ozone Depletion, Greenhouse Gases, and Climate Change, Proceedings of a Joint Symposium by the Board on Atmospheric Sciences and Climate and the Committee on Global Change, Commission on Physical Sciences, Mathematics, and Resources, National Academy Press, Washington, D.C., 1989.

Nierenberg, W. A., "Atmospheric CO_2: Causes, Effects, and Opinions," The Bridge (publication of the National Academy of Engineering), 18(3):4-11, Fall 1988.

Nowell, R. E., "Carbon Monoxide and the Burning Earth," Sci. Am., 261:(4), October 1989.

Oak Ridge National Laboratory, Energy Technology R&D: What Could Make a Difference? Part I: Synthesis Report, Contract DE-AC05-840R21400, U.S. Department of Energy, Washington, D.C., May 1989.

Oak Ridge National Laboratory, Environmental Sciences Division, Environmental, Health, and CFC Substitution Aspects of the Ozone Depletion Issue, Report ORNL-6552, U.S. Department of Energy, Washington, D.C., 1989.

Oak Ridge National Laboratory, Energy Efficiency: How Far Can We Go? Prepared for the Office of Policy, Planning and Analysis, U.S. Department of Energy, under Contract DE-AC05-84OR21400, Report ORNL/TM-11441, January 1990.

Oatman, P. A., D. J. Frey, and D. N. Wortman, Whole Building Energy Diagnostic System. Strategies for Reducing Natural Gas, Electric and Oil Costs, in Proceedings of the 12th World Energy Engineering Congress, Atlanta, Ga., 1989.

Odgen, J. M., and R. H. Williams, Solar Hydrogen - Moving Beyond Fossil Fuels, World Resources Institute, October 1989.

Pacific Northwest Laboratory, The Technology Transfer Process, White paper prepared for the Office of Policy, Planning, and Analysis, U.S. Department of Energy, April 1990.

Patrusky, B., "Dirtying the Infrared Window," MOSAIC (publication of the National Science Foundation), 19(3/4):25-37, Fall/Winter, 1988.

Peterson, J. L., J. W. Jones, and B. D. Hunn, "The Correlation of Annual Commercial Building Boil Energy with Envelope, Internal Load, and Climatic Parameters," ASHRAE Trans, Vol. 95, Part 1, 1989.

Ramanathan, V., et al., "Trace Gas Trends and Their Potential Role in Climate Change," J. Geophys. Res., 90:5547-55, 1985.

Rasmussen, K. "Sources, Sinks, and Seasonal Cycles of Atmospheric Methane", Geophys. Res, 88: 5131, 1983.

Ryan, W., J. Marsala, A. Lowenstein, and W. Griffiths, "Liquid Desiccant Residential Dehumidifier," in Proceedings of the 1989 International Gas Research Conference, Vol. II: Residential & Commercial Utilization, Atlanta, Ga., 1989.

Savitz, M., "The Federal Role in Conservation Research and Development," J. Byrne and D. Rich (eds.), The Politics of Energy Research and Development: Energy Policy Studies, Vol. 3, New Brunswick, N. J., Transaction Books, pp. 89-118.

Schmidt, E., "Sources and Sinks of Atmospheric Methane," Pure App. Geophys. 116:452, 1978.

Solar Energy Research Institute, The Potential of Renewable Energy, An Interlaboratory White Paper, Prepared for the Office of Policy, Planning, and Analysis, U.S. Department of Energy under Contract No. DE-AC02-83CH10093, Report SERI/TD-260-3674 DE90000322, March 1990.

Sperling, D., New Approaches to Transportation Fuels, University of California Press, Berkeley, Calif., 1988.

Sperling, D., and M. A. DeLuchi, "Transportation Energy Futures," Annual Review of Energy, 14:375-424, 1989.

Stein, R. G., C. Stein, M. Buckley, and M. Green, Handbook of Energy Use for Building Construction, Contract AC02-79CS20220, U.S. Department of Energy, Washington, D.C., 1981.

Turiel, I., and M. D. Levine, "Energy Efficient Refrigeration and Reduction of CFC Use," Annual Review of Energy, 14:173-204, 1989.

Tuft, P., and R. Norton, Energy Master Planning: Innovative Design and Energy Analysis Service (Ideas) for New Commercial Construction, Strategies for Reducing Natural Gas, Electric and Oil Costs, in Proceedings of the 12th World Energy Engineering Congress, 1989.

United Kingdom Department of Energy, Background Papers Relevant to the 1986 Appraisal of U.K. Energy Research, Development and Demonstration, ETSU-R-43, Reports compiled by Chief Scientists Group, Energy Technology Support Unit, Harwell Laboratory, 1987.

United Kingdom Department of Energy, Energy Technologies for the United Kingdom: 1986 Appraisal of Research, Development and Demonstration, Energy Paper 54, February 1987.

University of California, Lawrence Berkeley Laboratory, Effects of Low-Emissivity Glazings on Energy Use Patterns in Nonresidential Daylighted Buildings, Presented at 1987 ASHRAE Winter Meeting, Contract DE-AC03-76SF00098, Department of Energy, Washington, D.C., 1986, p. 132.

University of California, Lawrence Berkeley Laboratory et al., Energy Technology for Developing Countries: Issues for the U.S. National Energy Strategy, Draft Report, September 15, 1989.

U.S. Department of Commerce, Evaluating R&D and New Product Development Ventures: An Overview of Assessment Methods, PB86-110806, National Technical Information Service, Springfield, Va., 1986.

U.S. Department of Energy, Office of Renewable Energy Technologies, Five Year Research Plan, 1988-1992, Biofuels and Municipal Waste Technology Program, Washington, D.C., July 1988.

U.S. Department of Energy, A Primer on Greenhouse Gases: CO_2 Report No. DOE/NBB 0083, Office of Energy Research, Office of Basic Energy Services, Carbon Dioxide Division, Washington, D.C., March 1988.

U.S. Department of Energy, Energy Conservation Standards for Consumer Products: Dishwashers, Clothes Washer, Clothes Dryers, Technical Support Document CE/0267.

U.S. Department of Energy, Commercial Buildings Consumption and Expenditures, Report No. DOE/EIA-0318, Energy Information Administration, Office of Energy Markets and End Use, Washington, D.C., 1986.

U.S. Environmental Protection Agency, "Natural Gas: Can it Play a Major Role in Limiting Greenhouse Warming?", Global Climate Change Division, 1989, p.134.

Unnsach, S., et al., Comparing the Impacts of Different Transportation Fuels on the Greenhouse Effect, Acurex Corporation report to the California Energy Commission, April 1989.

Wacker, P. C., "Economizer Savings Study". ASHRAE Trans., Vol. 95, Part 1., 1989

World Meteorological Organization, Atmospheric Ozone 1985, Assessment of Our Understanding of the Processes Controlling Its Present Distribution and Change, Global Ozone Research and Monitoring Project, Report No. 16, Vol. III, 1985.

Williams, R. H., Biomass Energy Strategies for Coping with the Greenhouse Problems, Draft document, August 1989.

Williams, V. A., "Thermal Engineering Storage: Engineering an Integrated System. Strategies for Reducing Natural Gas, Electric and Oil Costs," in Proceedings of the 12th World Energy Engineering Congress, Atlanta, Ga., 1989.

ORDER CARD — CONFRONTING CLIMATE CHANGE

Use this card to order additional copies of **Confronting Climate Change** and the books described below. All orders must be prepaid. Please add $2.00 for shipping and handling. Prices apply only in the United States, Canada, and Mexico and are subject to change without notice. To order by phone using VISA/MasterCard/American Express, **call toll-free 1-800-624-6242**, Monday-Friday, 8:30-5:00 EST. Call (202) 334-3313 in the Washington metropolitan area.

____ I am enclosing a U.S. check or money order.

____ Please charge my VISA/MasterCard/American Express account.

Number: ______________________

Expiration date: ______________________

Signature: ______________________

Quantity Disounts:

5-24 copies	15%
25-499 copies	25%

To be eligible for a discount, all copies must be shipped and billed to one address.

PLEASE SEND ME:

Qty.	Code	Title	Price
____	CONCLI	Confronting Climate Change	$17.95
____	FUDRIV	Fuels to Drive Our Future	$24.50
____	OZOND	Ozone Depletion, Greenhouse Gases, and Climate Change	$20.00

Please print.

Name ______________________

Address ______________________

City ______________________ State __________ ZIP Code __________

CONC

Return this card with your payment to NATIONAL ACADEMY PRESS, 2101 Constitution Avenue, NW, Washington, D.C. 20418. Customers in Japan should send their orders to: Maruzen Co., Ltd., 3-10, Nihonbashi 2-Chome, Chuo-Ku, Tokyo 103, Japan. Customers in the United Kingdom, Europe, Africa, and the Middle East should send orders to: John Wiley & Sons, Ltd., 1 Oldlands Way, Southern Cross Trading Estate, Bognor Regis, West Sussex, P022 9SA, England.

OTHER BOOKS OF INTEREST:

FUELS TO DRIVE OUR FUTURE

This book explores the potential for producing liquid transportation fuels by enhanced oil recovery from existing reservoirs, by processing resources such as coal, oil shale, tar sands, and natural gas, and other promising approaches. **Fuels to Drive Our Future** draws together relevant geological, technical, economic, and environmental factors and recommends specific directions for U.S. research and development efforts on alternative fuels sources.

ISBN 0-309-04142-2; 1990, 240 pages, 6 x 9, index, hardbound, $24.50

OZONE DEPLETION, GREENHOUSE GASES, AND CLIMATE CHANGE

This volume presents the most up-to-date data and theories available on ozone depletion, greenhouse gases, and climatic change. These questions and more are addressed: What is the current understanding of the processes that destroy ozone in the atmosphere? What role do greenhouse gases play in ozone depletion?

ISBN 0-309-03945-2; 1989, 136 pages, 6 x 9, paperbound, $20.00